ACID RAIN AND THE RISE OF THE ENVIRONMENTAL CHEMIST IN NINETEENTH-CENTURY BRITAIN

T0304031

Science, Technology and Culture, 1700–1945

Series Editors

David M. Knight
University of Durham

and

Trevor Levere
University of Toronto

Science, Technology and Culture, 1700–1945 focuses on the social, cultural, industrial and economic contexts of science and technology from the 'scientific revolution' up to the Second World War. It explores the agricultural and industrial revolutions of the eighteenth century, the coffee-house culture of the Enlightenment, the spread of museums, botanic gardens and expositions in the nineteenth century, to the Franco-Prussian war of 1870, seen as a victory for German science. It also addresses the dependence of society on science and technology in the twentieth century.

Science, Technology and Culture, 1700–1945 addresses issues of the interaction of science, technology and culture in the period from 1700 to 1945, at the same time as including new research within the field of the history of science.

Also in the series

Matter and Method in the Long Chemical Revolution
Laws of Another Order
Victor D. Boantza

Selling Science in the Age of Newton
Advertising and the Commoditization of Knowledge
Jeffrey R. Wigelsworth

Meeting Places: Scientific Congresses and Urban Identity in Victorian Britain
Louise Miskell

Acid Rain and the Rise of the Environmental Chemist in Nineteenth-Century Britain

PETER REED

Independent Historian

Routledge
Taylor & Francis Group

LONDON AND NEW YORK

First published 2014 by Ashgate Publishing

Published 2016 by Routledge
2 Park Square, Milton Park, Abingdon, Oxfordshire OX14 4RN
711 Third Avenue, New York, NY 10017, USA

First issued in paperback 2016

Routledge is an imprint of the Taylor & Francis Group, an informa business

British Library Cataloguing in Publication Data
A catalogue record for this book is available from the British Library

The Library of Congress has cataloged the printed edition as follows:
Reed, Peter, 1942-
Acid rain and the Rise of the Environmental Chemist in Nineteenth-Century Britain / by Peter Reed.
 pages   cm. – (Science, Technology and Culture, 1700–1945)
Includes bibliographical references and index.
1. Smith, Robert Angus, 1817–1884. 2. Chemists – Great Britain – Biography.
3. Environmental chemistry – Great Britain – History – 19th century. 4. Acid rain –
Environmental aspects – Great Britain – History – 19th century. 5. Sanitation –
Research – Great Britain – History – 19th century. 6. Chemical industry –
Environmental aspects – Great Britain – History – 19th century. 7. Pollutants –
Government policy – Great Britain – History – 19th century. 8. Environmental health
– Great Britain – History – 19th century. 9. Great Britain – Environmental conditions
– History – 19th century. 10. Manchester (England) – Environmental conditions –
History – 19th century. I. Title.
QD22.S615R33 2014
577'.14092 – dc23
[B]   2013031539

ISBN 13: 978-1-138-24721-5 (pbk)
ISBN 13: 978-1-4094-5775-6 (hbk)

Contents

Contents

List of Figures

Permissions for Quotations

The author would like to thank the following publishers and authors for permission to quote from their books.

W.H. Brock, *Justus von Liebig, The Chemical Gatekeeper* (Cambridge, 1997). With permission of Cambridge University Press.

Alan Kidd, *Manchester: A History* (Lancaster, 2006). With permission of Carnegie Publishing Ltd.

Robert H. Kargon, *Science in Victorian Manchester* (Manchester, 1977). With permission of Robert Kargon.

Lynn McDonald (ed.), *Florence Nightingale on Society, Philosophy, Science, Education and Literature* (Waterloo, Ontario, 2003), Volume 5 in the *Collected Works of Florence Nightingale*, ed. Lynn McDonald. Used with the permission of Wilfrid Laurier University Press.

Tristram Hunt, *Building Jerusalem* (London, 2004). The Orion Publishing Group, London. Copyright ©2004 Tristram Hunt.

Tristram Hunt, *Building Jerusalem* (London, 2004). With permission of Capel & Land, Ltd, London.

Anthony S. Wohl, *Endangered Lives: Public Health in Victorian Britain* (London, 1974). With permission of Anthony Wohl.

Eric Ashby and Mary Anderson, *The Politics of Clean Air* (Oxford, 1981). By permission of Oxford University Press.

Mark Jackson, *Allergy: The History of a Modern Malady* (London, 2006). With permission of Reaktion.

Peter Reed, 'Robert Angus Smith and the Alkali Inspectorate', in E. Homburg, A.S. Travis and H.G. Schröter (eds), *The Chemical Industry in Europe, 1850–1914: Industrial Growth, Pollution, and Professionalization* (Dordrecht, 1998), pp. 149–63. With kind permission of Springer Science and Business Media B.V.

Abbreviations

BAAS   British Association for the Advancement of Science (renamed British Science Association in 2009)

BL   British Library

CC   Chadwick Collection, Special Collections Library, University College London

CRO   Cheshire and Chester Archives and Local Studies Centre

GMRO   Greater Manchester County Record Office

LRO   Liverpool Record Office

MLPS   Manchester Literary and Philosophical Society

MSSA   Manchester and Salford Sanitary Association

NA   National Archives

ODNB   *Oxford Dictionary of National Biography*

WCRO   Worcestershire County Record Office

Acknowledgements

Carrying out the research over quite a period of time and then writing a book such as this leads to the need to thank many people, institutions and organizations, without whose assistance the book would not have appeared.

It proved a major advantage to use two journal articles I wrote during the last few years as the basis of two of the chapters. I would like to thank the Newcomen Society and Maney Publishing for permission to use the article 'Acid Towers and the Control of Chemical Pollution 1823–1876' in *Transactions of the Newcomen Society* 78 (2008), pp. 99–126, and the Society for the History of Alchemy and Chemistry and Maney Publishing for permission to use 'The Alkali Inspectorate 1874–1906: Pressure for Wider and Tighter Pollution Regulation' in *Ambix* 59 (2012), pp. 131–51. I am grateful to several publishers and to Robert Kargan and Anthony Wohl for permission to quote from their books (see Permissions page).

The staff of many libraries and record offices provided unstinting support during all the stages by locating sources, manuscripts, books and articles, but especially during the research stage, and this has been a constant source of encouragement. I am especially indebted to the Librarian and staff of the London Library (and associated support from the Carlyle Trust), whose prompt service in dispatching books to my home saved much time and allowed the research and writing to continue with minimum delay. My special thanks go to Julie Ramwell and Janet Wallwork at the University of Manchester Library, who provided invaluable assistance in unlocking details of the Robert Angus Smith's library that was purchased after his death by a group of friends and then donated to Owens College before passing to the University of Manchester Library. The staff of the Greater Manchester County Record Office helped locate records of the Manchester and Salford Sanitary Association and the Manchester and Salford Noxious Vapours Abatement Association. Staff in the reference section of Manchester Central Library offered much help and advice during a period when services were relocated because of refurbishment of the Central Library building in St Peter's Square.

I am also indebted to David Allen and his staff at the Library of the Royal Society of Chemistry, who provided copies of several articles from the Society's publications and assisted with locating a number of images of well-known chemists. Bart Croes of the California Air Resources Board was very helpful in answering questions and providing information about the work of the Board and its first chairman, Arie Jan Haagen-Smit, when I visited their offices in Sacramento. Jennifer Harbster at the Library of Congress was very helpful in finding information on acid rain, particularly in relation to its impact on North America. Natalie Conboy at the Florence Nightingale Museum located correspondence between Florence Nightingale and Robert Angus Smith. Lutz Trautman at

Universitätsarchiv at Justus-Liebig-Universität, Giessen always provided a prompt response to my numerous enquiries about students who had studied with Liebig. Kate Forde at the Wellcome Institute, who was responsible for the excellent Dirt exhibition, provided in the early stages a very useful dialogue on issues raised by the exhibition as well as several sources of information. Paul Meara at the Catalyst Science Discovery Centre in Widnes was very helpful in answering queries and providing information.

Thanks are also due to the staff of the following libraries and record offices: British Library, Wellcome Library, National Archives, London Metropolitan Record Office, Bodleian Library, Special Collections Library at University College London, Society of Antiquaries of Scotland, National Library of Wales, National Library of Scotland, Birmingham Central Library, Leominster Library, Cheshire and Chester Archives and Local Studies Centre, and Gladstone's Library (Hawarden). I would like to thank a number of libraries and organizations for permission to use photographs from their collections (see Permissions page). My thanks also go to Derek Buick (Headland Design) for his work on the Leblanc process diagram.

I am also very grateful to the many people who, without realizing it, helped to shape this book by asking questions at the many conferences and seminars over the last few years when I spoke on topics related to the theme of the book.

Many individuals were supportive at different stages, sharing information, answering questions or reading draft chapters of the book. I would like in particular to thank Richard Noakes, Rebecca Whyte, John Hudson, Mike Emanuel, Robin Betteridge, Frank James, Bill Brock, John Perkins, Thomas Le Roux, Bill Griffith and Ernst Homburg.

I would especially like to acknowledge the support of Tony Travis, from the early stages, for his willingness to read through many draft chapters and for his detailed comments that have improved the structure and contents of the book. Nevertheless, as always, any errors or omissions remain solely my responsibility.

I would like to thank Ann Sharp, who over many months brought order to my research by deciphering and typing my hand-written notes made during visits to numerous libraries and record offices; this allowed the research and writing to proceed with more purpose.

At Ashgate Publishing, Emily Yates has provided constant support and has remained a steady hand during the writing and production stages, always providing prompt answers to the many queries, especially during the final editing stages when progress is dependent on speedy responses, Kirsten Weissenberg has seen the book through the final stages of editing and production and Emma Gallon has provided useful guidance on images. My thanks also go to Elizabeth Teague for her diligent proofreading. I am also grateful to David Knight and Trevor Levere as editors of the 'Science, Technology and Culture, 1700–1945' series, for recognizing the value of a book on the life and work of Robert Angus Smith in the context of his work on acid rain and his roles as environmental chemist and government inspector.

My warmest thanks go to Margaret Anne Olmsted, who, over a long period of time, was drawn into regular discussion on themes and topics without knowing how they would all fit together finally in the book. Her interest and encouragement throughout the long process, and her assistance with getting permission to quote from many books, have been crucial. I hope the published book provides a satisfactory answer and is justification for all her time and efforts.

PETER REED
Leominster
April 2013

Chapter 1
Introduction: Short Biography and Emergence of the Civil Scientist

Short Biography

Robert Angus Smith was born on 15 February 1817 in Pollockshaws near Glasgow, the seventh son and twelfth child of John and Janet Smith.[1] John was a minor manufacturer in Loudoun, Ayrshire and in 1796 married Janet Thomson, daughter of a flax-mill owner, at Avondale, Lanarkshire.

Pollockshaws, located a few miles to the south of the centre of Glasgow, was quite a thriving commercial and manufacturing centre in the early 1800s; it had the first bleach fields and print works in the west of Scotland and these were followed by businesses in linen manufacture, brewing, dyeing, engineering, paper manufacture and pottery.[2] Even in this thriving economic environment the Smith family failed to prosper and life proved a constant struggle and a battle with poverty; perhaps this hardship is not surprising given the family's large size, but it was probably due more to John's lack of business acumen, and by 1820 the Smith family was increasingly dependent on the sons for its financial survival.

While always relatively poor, the family was dominated by John's fierce religious fervour, which was to have a profound influence on members of the family, and particularly on Robert as he struggled in his early years to find fulfilment and a direction to his life. Day-to-day family life was dictated by strict religious observance of Calvinism, with 'silence enforced at mealtimes and

---

[1] For biographical information, see E. Schunck, 'Memoir of Robert Angus Smith', *Memoirs of the MLPS*, 10 (3rd series) (1887), pp. 90–102; William Anderson Smith, *Shepherd Smith, The Universalist* (London, 1892); Christopher Hamlin, 'Entry for Robert Angus Smith', *ODNB*; Peter Reed, 'Entry for Robert Angus Smith', *Biographical Dictionary of 19th Century British Scientists* (Bristol, 2004), pp. 1843–6; Alan Gibson, 'Robert Angus Smith and Sanitary Science', MSc. Diss., University of Manchester Institute of Science and Technology, 1972; 'Robert Angus Smith, M.D., F.R.S.', *The British Medical Journal*, 1 (1884), p. 976; 'Robert Angus Smith', *Journal of Chemical Society*, 47 (1885), pp. 335–7; 'Obituary: Robert Angus Smith, LL.D, F.R.S.', *Journal of the Society of Chemical Industry* 3 (1884), pp. 316–17; 'Obituary: Robert Angus Smith', *American Journal of Science*, 128 (1884), pp. 79–80.

[2] Jack Gibson, *Pollockshaws: A Brief History* (Pollockshaws Heritage Group, 2006). In 1807 one of the cotton mills was lit by gas produced within the mill, the first in Scotland.

the discouragement of unnecessary talking'.[3] One of John's sons, James, later complained that his father had 'spoiled the language of the family',[4] and wrote of the effect on his brother, Joseph:

> He is uncommonly ignorant of all the common realities of life. I wanted to take him to the theatre, but he objected. He had never read a play or novel in his life; and I question if he ever even read a history or a newspaper. This comes of strict religious notions.[5]

As if this was not difficult enough, the family was later shaken further by its fervent support for the schism created by the theologian Edward Irving with its prophetic speculations that Christ would soon return.[6] This was one of many schisms to shake religious belief during the first half of the nineteenth century, but the Irving schism was to have a lasting influence on the lives of several members of the Smith family.

While sharing her husband's religious devotion and wanting her sons to enter the Church, Janet had a very different temperament from her husband, providing the calm and affectionate counterpoint to his strict religious discipline. As one grandson recalled, she had

> the charm of calm content in her large, soft Scotch face – the content that comes from a mind and body incapable of the pettiness of fretting, that has lived a life of duty and devotion.[7]

Alongside the religious fervour, there was a commitment to providing the sons with a good education and inculcating belief in hard work and integrity. Although there were 13 children, information has survived about only five of the seven sons. The eldest, John (1800–1871), graduated from Glasgow University in 1819 with an MA. Although John had expected to enter the Church, he became a schoolteacher instead, having found himself in the more pressing role of having to support the family financially after his father stopped working. After a period teaching at Lauriston Academy, Glasgow, John moved to Perth Academy, where he taught mathematics before his appointment as headmaster of the school. He proved an outstanding educationalist and in 1854 was elected President of the Educational Institute of Scotland. The Institute was founded in 1847 and is today the oldest

---

[3]   Gibson, 'Angus Smith', p. 1.1.

[4]   Ibid., p. 1.2.

[5]   Ibid.

[6]   David Bebbington, 'Gospel and Culture in Victorian Nonconformity', in Jane Shaw and Alan Kreider (eds), *Culture and the Nonconformist Tradition* (Cardiff, 1999), p. 56. It is interesting to note that in 1821 Irving fell deeply in love with Jane Welsh, the future wife of Thomas Carlyle.

[7]   Smith, *Shepherd Smith*, p. 18.

teachers' trade union in the world.[8] From a young age John showed a keen interest in science, an interest he retained throughout his life. Later his scientific papers on optical experiments were published in the *Proceedings* and *Memoirs of the Manchester Literary and Philosophical Society*, no doubt with the support and encouragement of his brother Robert, who was an officer of the Society.[9]

The second son was James Elishama (1801–57), later known as 'Shepherd Smith'. James too graduated with an MA from Glasgow University (1818) and although he was ordained, he never became a minister of the Church of Scotland. It was James who disrupted the Smith family's religious sensibility in 1828, for having heard Edward Irving preach during a tour of Scotland, he converted many of the family to Irvingism. James's father was so influenced by Irvingism that he became a lay preacher in the sect, while James would later abandon the sect and come under the influence of John Wroe, a follower of the 'Christian Israelites' sect in Ashton-under-Lyne through which he developed friendships by correspondence with a number of women, most notably with Rosina Lytton, wife of the writer and politician, Edward Bulwer-Lytton (later Baron Dalling and Bulwer). On James's death Robert, with moral indignation, destroyed this collection of letters because of

> The atrocity of their revelations and the fear that they should one day be made public. He consequently used afterwards to declare that no one held Lord Bulwer Lytton in more absolute contempt morally than he did; and yet no one has done more for his reputation.[10]

Another son, Micaiah, also graduated with an MA from Glasgow University and became a noted Arabic scholar. Early in his life he was ordained, like James, but never became a minister of the Church. Micaiah was a leading promoter of Irvingism and remained a loyal member of the sect, acting as tutor to many families sharing the Irving faith; for a short time he was tutor to the family of the British consul in Tangiers.

The only other brother about whom any information survives is Robert's younger brother, Joseph. He studied chemistry under Professor Frederick Penny at the University of Aberdeen and it is quite likely that he and John aroused and nurtured Robert's early interest in the subject through their regular correspondence and study of the standard chemistry texts of the time, many of which were in French.

As his brothers, Robert was expected from a young age to pursue a career in the Church, and his education was aimed towards that goal, but the expectation weighed heavily on his mind as he searched and pondered the future direction

---

8   The Educational Institute of Scotland was granted a Royal Charter by Queen Victoria and is the only trade union able to award degrees.

9   John Smith, 'Cause of Colour', *Proceedings of the MLPS*, 1 (1860), pp. 147–9. Also John Smith, 'On the Origin of Colour and the Theory of Light', *Memoirs of the MLPS*, 1 (3rd series) (1862), pp. 1–96, and *Proceedings of the MLPS*, 7 (1868), pp. 137–40.

10   Smith, *Shepherd Smith*, p. 325.

of his life (as many other young people of the time and since). This dilemma was to leave him uncommitted and wavering for very many years. However, what was not in dispute was the commitment of both parents to provide the best education for their sons, and, aged nine, Robert started attending Glasgow High School to study the classics, before following in his brothers' footsteps to Glasgow University in 1828 where he pursued his classical studies. It was felt at the time (and remained an educational principle for many well into the twentieth century) that studying the classics created the foundation of knowledge and understanding upon which all other subjects could be added. Unfortunately, Robert was forced to leave the University the following year without completing his degree and this added further uncertainty to the future direction of his life.[11] While there are many possible reasons, the most likely is the family's continuing financial hardship, exacerbated when John, Robert's brother, who had become the main breadwinner, married and left home in the same year.

Without the clear direction a completed university course might have provided, Robert was left floundering, unable to make up his mind whether to become a minister in the Church or pursue another career opportunity. But his short time at university brought a lifelong friendship with James Young, the chemist and industrialist, whom Robert met while attending the chemistry lectures given by Thomas Graham at Anderson's University in Glasgow.[12] When Graham was appointed Professor of Chemistry at Anderson's University in September 1830, his arrival generated a good deal of public interest. Many were drawn to his series of lectures and Robert attended 'one or two of his popular lectures'.[13] While this cannot be taken as firm conversion to a career in chemistry, it was subsequently to prove very influential for Robert in two ways – a lasting friendship with Graham, who became a mentor, and the further blossoming of the interest in chemistry initiated by his brothers. Even so, it is important to remember that at this time Robert was only 13 years of age, hardly the age for making such life-determining decisions.

Leaving university in his rudderless state of mind, and to allow time for further reflection about his future, Robert became tutor to a family in the Scottish Highlands, the first of several such roles. Many of the families for whom he was tutor shared the Irvingate philosophy and would have been reassured that their children had a tutor with sympathies for their religious beliefs. It would be much easier for Robert to contact these families through the Irvingate network. His movements after leaving university are not well recorded but we know that he arrived in London in 1836 to stay with his brother Micaiah, and while there he is

---

[11]　A. Gibson and W.V. Farrar, 'Robert Angus Smith, F.R.S., and "Sanitary Science"', *Notes and Records of the Royal Society of London*, 74 (1974), p. 241.

[12]　Some historians have suggested that Robert met Lyon Playfair and David Livingstone at the University of Glasgow or at Anderson's University. Playfair and Livingstone were in Glasgow much later, when Angus Smith was in London.

[13]　Robert Angus Smith, *Life and Works of Thomas Graham* (Glasgow, 1884), p. 65.

likely to have retained his tutoring role. He is also known to have spent time with another brother, James, who records:

> Poor Robert seems to have been brought up in a balloon or a coal-pit, or some place out of the world altogether. I advised him to endeavour to pick him a little more information respecting the daily occurrences of society, but whether he follows my advice or not I know not.[14]

This is quite a disturbing statement about a 19-year-old, even making allowance for his difficult home life and his sudden exit from university, and further confirms his social isolation, compounded by the continuing uncertainty over his future career. Even two years on, the general malaise continued:

> Robert is strong studying Greek, and knows nothing about the world and its concerns. He is living in the first or second century, with a sort of prophetic glimpse into the nineteenth. But if it do him no harm in a pecuniary point of view, he is quite happy in one century as another.[15]

Robert continued with his tutoring work and in 1837 was tutoring the family of the Rev. and Hon. H.E. Bridgeman, and two years later accompanied them to Germany. When the family returned to Britain, Robert remained in Germany, and without realizing it at the time, started down a path that would eventually change his life and bring outstanding achievement and success. In February 1840 he enrolled at Giessen University to study under the renowned chemist, Justus Liebig. But what had caused Robert to register as a student at Giessen?

Liebig was an outstanding chemist who had created a chemistry course at Giessen that was attracting students from all over the world and especially from Britain, where there were no similar institutions. The method of training encompassed learning the fundamental principles of chemistry and then applying them in special research projects under Liebig's personal mentoring. There is no clear evidence about whom Robert consulted or whether indeed he consulted anyone about Liebig's course at Giessen. It may be that he consulted with Graham, who was well aware of the standing of Liebig and his course, since Liebig was already having a major influence on science in Britain. Hearing of Liebig and Giessen while in Germany may have been a serendipitous event. Nevertheless, registering as a student in such a renowned institution further confirms Robert's increasing fascination with chemistry at this period in his life.

Little is known of Robert's time in Giessen except that he registered in February 1840 and attended the course with Liebig until he matriculated in February 1841. He was awarded a PhD on 7 May 1841 without submitting a doctoral thesis. This was not an infrequent occurrence at Giessen, where Liebig often took it upon

---

14   Smith, *Shepherd Smith*, p. 150.
15   Ibid., p. 193.

himself to award degrees in such circumstances.[16] Besides the opportunity to study chemistry, a further advantage of coming under Liebig's influence was his promotion of the idea of applying chemical knowledge and understanding for the benefit of mankind, with particular emphasis on physiology and agriculture.

> The higher economic or material interests of a country, the increased and more profitable production of food for man and animals, as well as the preservation and restoration of health, are most closely linked with the advancement and diffusion of the natural sciences, especially chemistry.[17]

This role for chemists alongside advancing chemical knowledge and understanding probably engaged, perhaps even inspired, Robert and was to determine the future direction of his life. Another benefit of the Giessen course was the friendships nurtured among the students; these friendships were valuable during the course, but they proved even more important in the years that followed since a network was formed of those who were to go on advancing chemical understanding and also become leaders in major institutions around the world. For Robert, the two fellow students who were subsequently to play such an influential part in his life were Lyon Playfair and Edward Schunck.[18]

After finishing the course in Giessen, Robert decided to remain in Germany for a short time to improve his knowledge of the German language and his appreciation of the culture before returning to London. Over the period 1841 to 1843 Robert was again in a quandary, as he had been before leaving for Germany, but he continued his tutoring and assisted his brother James with editing his magazine, *Family Herald*. His mind was restless again; he still did not know whether to pursue a career in chemistry, a subject that had fascinated him during his studies in Giessen, or become a minister in the Church. Robert's indecision may reflect the guilt he felt over letting his parents down, especially since several of his brothers had chosen the Church. He had by this time firmly distanced himself from Calvinism and the Church of Scotland, and was now attracted by the Anglican Church, but without an appropriate university degree he was barred from becoming a minister. It is not known whether Robert consulted others again about this dilemma, but matters came to a head in 1843 when Lyon Playfair, a fellow student in Giessen, and having been appointed Professor of Chemistry at the Royal Manchester Institution earlier in the year, invited him to become his assistant.

Robert appears to have readily accepted Playfair's invitation to join him in Manchester, and this decision forced Robert's hand in the direction of a future

---

[16]   I am grateful to Lutz Trautmann (Universitätsarchiv at Justus-Liebig-Universität, Giessen) and Bill Brock for this information.

[17]   Justus von Liebig, *Familiar Letters on Chemistry*, 3rd edn (London, 1851), p. 19.

[18]   Armin Wankmüller, 'Ausländische Studierende der Pharmazie und Chemie bei Liebig in Giessen', *Deutsche Apotheker-Zietung*, 107/14 (1967), pp. 463–7. See also T.E. James, 'Entry for Edward Schunck', *ODNB*.

Figure 1.1   Robert Angus Smith. The photograph was probably taken between
              1857 when Smith was elected a Fellow of the Royal Society and
              1864 when he was appointed Inspector under the Alkali Act.
              Reproduced courtesy of the Library of The Royal Society of
              Chemistry

career in chemistry, although the exact career path would remain unclear for several more years. No records have survived to show whether Robert agonized further about entering the Church. The post of Professor of Chemistry at the Royal Manchester Institution was unpaid, although the holder of the post was provided with rooms he could fit out as a laboratory. The main income was student fees from the courses and, since they were fortunately well attended, Playfair was able to pay his assistants a small remuneration and offer some laboratory space. The influence of Graham, Liebig and Playfair on Robert Angus Smith is discussed in more detail in Chapter 2.

Robert settled readily into life in Manchester, which by the 1840s had become a city that epitomized the effects of mass industrialization, both the benefits of work, mass employment and prosperity; and the adversities of poverty, poor housing and insanitary living conditions. In 1843, following Edwin Chadwick's report, *The Sanitary Condition of the Labouring Population*, the Royal Commission on the Health of Towns was set up to review conditions in towns across Britain. Playfair was appointed one of the commissioners and took as his special study the large towns of Lancashire, of which Manchester was the most important, and immediately asked Robert to join him as an assistant commissioner, a post that Robert readily accepted. By 1845, work for the Commission was complete and Lyon Playfair moved to London to take up the post of Chemist at the Geological Survey, while Robert was left in another quandary about his future. By this time, though, he was well settled in Manchester and, with his work as a consulting and analytical chemist going well, he decided to stay there.

Work for the Royal Commission brought Robert into close proximity with the harsh insanitary conditions then prevailing in the large manufacturing towns, and such experiences were to instil in him a lifelong interest in air and water quality and in disinfection. The following year, as he walked into the city one morning, he was so appalled by a pall of black smoke hanging over Manchester, giving rise in him to an almost apocalyptic experience, that he wrote to the editor of the *Manchester Guardian*, in a vein Ruskin would have empathized with, that the general blight of mass industry on the environment was in direct conflict with God's creative work.[19] This experience probably triggered Robert's lifelong experimental interest in air quality that would embrace investigations covering the contents of the air under differing circumstances, at different places in Britain and abroad, and in different climatic regions, in an attempt to ascertain how the different agents in the air affect its quality as well as the air's influence on the environment and on people's health.

In 1857, Robert agreed to assist his friend, Peter Spence, the alum manufacturer at Pendleton, near Manchester, who was facing accusations that his works caused a nuisance because of the various gases they were allegedly releasing into the atmosphere. Despite determined efforts by Robert and others, who surveyed the works on several occasions, a prosecution was mounted and Spence lost the case.

---

[19]   *Manchester Guardian*, 2 November 1844 and 20 November 1844.

Subsequently he had to relocate the works. But for Robert the more disturbing aspects of the trial were the use of chemical consultants as expert witnesses, the manner in which the expert scientific evidence was treated by the court, and how expert evidence and witnesses might be bought by the highest bidder. Robert was deeply affected by the experience, and over many years he set in motion a campaign that embraced newspapers, journals, the Royal Society of Arts, the Chemical Society and the British Association for the Advancement of Science. This courtroom experience greatly influenced the manner in which he conducted his later responsibilities with the Alkali Inspectorate when legal challenge was the government's preferred method of persuading industry to comply with the terms of parliamentary regulations. In 1857 Robert Angus Smith was elected a Fellow of the Royal Society, with the support of Thomas Graham, Wilhelm Hofmann, Lyon Playfair, Edward Frankland, Edward Schunck and John Tyndall for his work on air and water quality.[20]

In 1859, in his article *On the Air of Towns*, Angus Smith became the first person to use the term 'acid rain' in referring to rain with an acid characteristic due to containing sulphur dioxide from the burning of coal with high levels of sulphur.[21] As the foremost authority on air quality, Angus Smith carried out a number of investigations for the Royal Commission on Mines (1864). His work on air quality and 'acid rain', and his experience of the Spence court case, are discussed in more detail in Chapter 4.

With his wide interests and his working base established in Manchester, Angus Smith applied for membership of the Manchester Literary and Philosophical Society (MLPS) and was elected on 29 April 1845, no doubt with encouragement from Lyon Playfair, who had became a member in 1842.[22] The Society, its meetings and publications were to become the intellectual centre for Angus Smith in preference to similar societies in London. Besides his prolific contributions to the meetings and publications (as his personal bibliography shows), he took more than a reasonable share of administrating the Society. He was Secretary between 1852 and 1856 (jointly with Edward Schunck in 1856), Vice-President over the period 1859 to 1878 (often jointly with Schunck) and President from 1864 to 1865. His major publications for the Society include *Memoir of Dr Dalton, and History of Atomic Theory* (1856) and *A Century of Science in Manchester* (1883), a review of the contribution of the Manchester Literary and Philosophical Society.

---

[20]   *Certificate for Election of Candidate for Robert Angus Smith*, 22 January 1856, Royal Society Library.

[21]   Robert Angus Smith, 'On the Air of Towns', *Quarterly Journal of the Chemical Society*, X1 (1859), pp. 196–235. It is often assumed the term was first used in the 1970s and 1980s when acid rain was causing major damage across several continents.

[22]   Lyon Playfair was elected a member of MLPS on 25 January 1842 and was elected an honorary member on 29 April 1851. It is interesting to note that Thomas Graham and A.W. Hofmann were elected honorary members on the same date, 23 January 1866, when Angus Smith was Vice-President of the Society.

The MLPS drew together a remarkable group of fellow scientists and engineers that included James Young (Angus Smith's friend from Glasgow days), Frederick Crace-Calvert (a fellow consulting chemist who took over from Playfair at the Royal Manchester Institution and worked on the disinfectant properties of carbolic acid), Edward Schunck (a fellow student at Giessen), Peter Spence (the alum manufacturer), Edward Frankland (who became Professor of Chemistry at Owens College) and Alexander McDougall (who patented a disinfectant powder with Angus Smith) – all were to play important roles in Angus Smith's professional work over the next 30 or so years.

During the 1840s, concerns that sanitary conditions were steadily getting worse were expressed by the local authorities in Manchester and Salford, local churches, medical officers of health, organizations such as the MLPS and many of the prominent citizens. An important initiative lead by the nexus of sanitary scientists at the MLPS was the foundation in 1852 of the Manchester and Salford Sanitary Association. One of the main objectives of the Association was to provide talks and information leaflets for people in their local communities and demonstrate how personal effort could improve their own sanitary circumstances. Angus Smith was an active supporter of this work, giving talks in church and community halls in an effort to improve sanitation and health. A fuller account of Angus Smith's life in Manchester is provided in Chapter 3.

From his work on air quality, Angus Smith was well aware of the pollution caused by alkali works using the Leblanc process due to the hydrogen chloride gas (or muriatic acid gas, as it was called in the alkali trade) released into the atmosphere.[23] This process had been adopted in Britain from about 1814 to produce sodium carbonate from salt rather than relying on the increasingly inadequate natural sources of alkali, which were barilla, a plant grown extensively around Alicante, and kelp, a *focus* genus of seaweed, found around the Scottish and Irish coasts. The gas was released from tall chimneys in an attempt to disperse it, but this proved ineffectual, with the hydrogen chloride descending either as a gas or as a liquid acid, causing damage to the surrounding area by affecting not only the natural environment but also local people and their homes. Legal claims for damages proved difficult to achieve within the existing legal framework, but when the wealthy landowners found their land and woodlands badly affected and financial values severely reduced, it was not long before parliamentary action was under way. Lord Derby, whose Knowsley Hall estate near Liverpool was affected by the pollution from St Helens, moved in 1862 to have the House of Lords set up a Select Committee on Injury from Noxious Vapours (dubbed the 'Smells Committee' by *Punch*).[24] During evidence to the Select Committee,

---

[23]   The word 'pollution' was not used to denote environmental degradation until 1870; see 'Air Pollution by Chemical Works', *Quarterly Journal of Science*, 7 (1870), pp. 330–41. For general discussion, see Peter J. Thorsheim, PhD diss., University of Wisconsin–Madison, 2000, pp. 7–8.

[24]   *Punch*, 24 May 1862, p. 204.

William Gossage, an alkali manufacturer from Worcestershire, revealed details of the tower he had invented (his acid tower) in which the hydrogen chloride gas was condensed in water with almost 100 per cent success.[25] The acid tower provided the basis for the Alkali Act of 1863 that required alkali works to condense at least 95 per cent of their hydrogen chloride gas, but the legislation (under the Board of Trade) also demanded unconditional access to the alkali works for checking the degree of condensation, as well as provision for an inspector to ensure that the regulations were enforced.[26] In February 1864 Robert Angus Smith was appointed the first Inspector (later redesignated as Chief Inspector).[27]

From his base in Manchester, Angus Smith set about organizing the team of sub-inspectors needed to cover the different parts of the country and began to draw up the procedures for inspecting the alkali works in enforcing the regulations. Because releases of gas were most likely overnight and under cover of darkness, quite sophisticated aspirators were devised to collect samples of gas in the flues at regular intervals over several days, and the receptacles were sealed to avoid any interference. Gaining access to the works required the cooperation of the manufacturers; initially, many were lukewarm, and some even hostile, but gradually Angus Smith's cooperative approach, together with the sub-inspectors acting as peripatetic consultants, allayed the manufacturers' fears about unreasonable interference in the operation of their plant and damage to their commercial prosperity. In 1872, as part of the Public Health Act, responsibility for the Alkali Inspectorate (also sometimes referred to as the Alkali Administration) passed from the Board of Trade to the Local Government Board that was also responsible for Poor Law administration and public health. With many industries expanding and diversifying rapidly, and several releasing nuisance gases into the atmosphere, there was a constant need to add to the list of industries and processes requiring inspection and regulation. Both the general public and diverse campaign groups kept up pressure on Angus Smith and the Alkali Inspectorate for better enforcement of the existing regulations since large quantities of objectionable gases continued to be released into the atmosphere. This ensured that unregulated works causing pollution did not escape scrutiny.

The 1863 legislation, and its later revision in 1868 and 1872, was to protect property, but from 1874 much more of the focus was directed at the health issues for industrial workers and for people living in close proximity to the noxious gases. When the Local Government Board took over responsibility for the Alkali Inspectorate there was an expectation that Angus Smith and John Simon (Medical Officer) might work more closely to address these issues from within the same department, but Simon became so preoccupied with his conflict with the Permanent

---

[25]   *British Patent* 7267/36. For Gossage's evidence to the Select Committee, see *Report of the Select Committee on Injury from Noxious Vapours* (486), P.P. 1862, XIV, pp. 82–93.

[26]   *Alkali Act 1863*, P.P. 1863 (220), pp. 1–2.

[27]   *Chemical News*, 10 (6 February 1864), p. 72.

Secretary, John Lambert, that the benefits never materialized and progress proved very slow. Nevertheless, regular reports to Parliament and a number of publications highlighting the dangers inherent in industrial workplaces led to the appointment of a Chief Inspector of Factories in 1896 and a Medical Inspector of Factories in 1898. The evolution of pollution control and associated health concerns during Angus Smith's time with the Alkali Inspectorate are discussed in Chapters 5, 6 and 7.

From his early days in Manchester, Angus Smith was shocked by the insanitary conditions in the city, conditions that Engels and Dickens described so vividly in *The Condition of the Working Class in England* and *Hard Times* respectively. The experience of these conditions had set in motion Angus Smith's preoccupation with investigating air and water quality, identifying their foreign constituents and their origins, so that the effects on people were better understood and living and working conditions were improved. His work with Lyon Playfair for the Royal Commission on the Health of Towns and his work on behalf of the Manchester and Salford Sanitary Association with local communities gave Angus Smith a vivid insight into the scale of the sanitary problems to be resolved, but the lack of scientific understanding of the agents at work undermined any major progress. As Christopher Hamlin has pointed out:

> He worried also about the roles of microscopic life in causing disease and decomposition, and also in completing the process of the decomposition of organic wastes. Smith was one of the few who recognized that Liebig's chemical theory of decomposition-based pathology was not necessarily irreconcilable with Louis Pasteur's biological theory. He became interested not simply in whether microbe species caused disease but in the chemical ecology of microbial life, in the environmental factors that either generated pathogenic microbes or harboured and cultivated them between human hosts.[28]

It was in relation to this work that Angus Smith worked with Alexander McDougall, a fellow member of the MLPS, to develop a disinfectant powder that was patented in both names in 1854.[29] Angus Smith seems to have done little of the promotion and marketing of the powder; this was left to McDougall.[30]

Many of Angus Smith's talks and articles for the MLPS addressed issues such as lead in water, sewage and sewage rivers, the development of living germs in water

---

[28]   Hamlin, 'Angus Smith'.

[29]   B.P. 1854/142, *Deodorizing and Disinfecting Sewage Matters, and Separating Manure therefrom.*

[30]   Angus Smith did not apparently gain financially from the disinfectant powder. For information on the McDougall family and their flour-making business, see Ellen McDougall, *The McDougall Brothers and Sisters* (London, 1923) and the centenary brochure, *A Matter of History: McDougalls* (1964).

and organic matter in water.[31] Angus Smith brought together much of his research and current knowledge and understanding in 1869, when he published *Disinfectants and Disinfection*, which was perhaps one of his most successful books.[32]

Florence Nightingale was a regular correspondent with Angus Smith on general public health matters, and in 1861 she asked him to write a report on ways of improving water sanitation in several towns of India as part of her work there. Nightingale then had the report printed at her own expense and distributed to designated officials in India who were involved in sanitary administration.[33] Angus Smith's work on disinfectants and the cause of disease is discussed in more detail in Chapter 4.

From the mid-1850s onwards there were major concerns about the increasing pollution of some rivers in Britain. The country had changed markedly with the urbanization and mass industrialization. One area of the country causing alarm was the north-west of England, with the major river systems of the Irwell and Mersey. The Irwell basin included the towns of Manchester and Salford as well as Bolton, Rochdale and Bury, while the Mersey basin served south Lancashire and north Cheshire. Although the major polluter of these rivers was untreated human sewage, industry made a huge contribution, with dye companies, paperworks, cotton mills, tanneries and alkali works among the main culprits.[34]

Many of the industrialists giving evidence to the 1876 Royal Commission on the Best Means of Preventing Pollution of Rivers (Mersey and Irwell Basin) accepted that the river systems were badly polluted, but were ready to point an accusing finger at the town sewage rather than accept liability for the effluent from their own works. There was general recognition of the sources of pollution, but the situation would prove very difficult to remedy in the short term: town sewage treatment required improved means of collection as well as construction of large treatment plants; industrial effluent required careful treatment on site before the water was discharged into the rivers, streams or canals; and most industrial sites had totally inadequate land area on which to construct treatment plants. While swift progress was expected following the Royal Commission's report, major improvements to water quality would take much longer to achieve.[35]

---

[31]   These include: 'Lead in Manchester Water', *Proceedings of MLPS*, 2 (1862), pp. 30–32; 'On Sewage and Sewage Rivers', *Memoirs of MLPS*, 12 (2nd series) (1855), pp. 155–76; 'On the Development of Living Germs in Water', *Proceedings of the MLPS*, 22 (1883), pp. 25–32; 'On the Examination of Water for Organic Matter', *Proceedings of MLPS*, 7 (1868), pp. 72, 78–9. Angus Smith also developed a coating to protect iron pipes.

[32]   Robert Angus Smith, *Disinfectants and Disinfection* (Edinburgh, 1869).

[33]   Letter: Florence Nightingale to Sir Harry Verney, 9 June 1861. WL (Ref: Ms 8999/23).

[34]   Gerard Francis Gordon, *New Perspectives on River Pollution Control in the Mersey and Irwell Basin (1868–1900)*, MSc Diss., University of Liverpool, 1993, pp. 1–5.

[35]   Bill Luckin, *Pollution and Control* (Bristol, 1986), pp. 164–7.

As with the updated Alkali Acts, the Rivers Pollution Act of 1876 stipulated that treating industrial effluents and urban sewage before discharge required the best practical means. With Angus Smith's investigations on town sanitation and sewage, and with his success in the Alkali Inspectorate working with industrialists to control pollution and enforce regulations, few were surprised when he was appointed as Inspector to work alongside the existing Inspector, Robert Rawlinson. This appointment as he approached the age of 60 was a major undertaking for Angus Smith, who was already carrying huge responsibility for the Alkali Inspectorate during a period when many legislative changes added to its responsibilities and accountability. With little time for relaxation and for pursuing his personal interests, it is very likely that work pressure through the 1880s from the dual appointments began to affect his health adversely. Nevertheless, displaying his usual fortitude, he remained resolutely determined to continue fulfilling the responsibilities associated with the two government inspector posts.

Perhaps borne out of his childhood and his classical education, Angus Smith retained a restless mind, always open to an enquiry engaging his attention. This is shown most clearly in the range of topics forming his numerous contributions to the MLPS. While most papers are addressed to scientific and technical matters, as might be expected, many others show his wider interests, with papers on such diverse topics as: 'On the history of the word "chemistry" or "chemia"'; 'The eucalyptus near Rome'; 'On vitrified forts'; 'Ancient maps of Africa'; 'On a remarkable fog in Iceland'. Some of these papers are quite short while others explore topics in great depth, but they all draw on thorough research, whether investigations in the field or in the laboratory, followed by detailed reading and study.

In about 1867 and quite late in his life, Angus Smith began to take a serious interest in antiquarian topics. This may have coincided with a holiday in Oban, Scotland, when he was persuaded by friends to visit the vitrified fort at Dun MacUisneachan, a few miles north of Oban. Such forts occur in a number of locations in Scotland, and the stone used in their construction carried a glazed surface as if it had been treated in some way at high temperature. This surface treatment made the stone resistant to air and helped its preservation. During his first visit, intrigued by the scientific aspects of the glazed surface, Angus Smith resolved to return to study the forts in even more detail. However, it was not until 1869 that he was able to take time off from his duties at the Alkali Inspectorate to devote to his investigations. The results were included in a paper for the Society of Antiquarians of Scotland in 1870 and published in book form later the same year.[36] His investigations were to continue over a number of years and in 1882 he was visiting an area near Fort William to investigate a further glazed-stone fort close to Ben Nevis:

---

[36]   Robert Angus Smith, *Descriptive List of Antiquaries near Loch Etive* (Glasgow, 1870).

The name of this fort, along with two or more of the same kind, is Dun Deardhuil, which the Rev. Dr. Clerk, of Kilmallie, says may mean 'Fort of the shining eye'. This name anglicised by Macpherson into Darthula is that given to the great beauty of the western celts, the Ulster Lady, called in Ireland Deirdre. She was the 'Helen' of the race, and her history and its indirect consequences brought out the Iliad of Ireland ... The similarity of name and the fact that Deirdre came over to Loch Etive may have caused people to connect these forts with her, but the fact of their being places of security requiring also vigilance is a reason for connecting them with the 'shining eye', a name, too, that may be easily supposed applicable to Deirdre. She was vigilant and beautiful. They were vigilant, but not beautiful. The work of the forts is rough; shapeless and small stones are melted together by a rude glaze. All known to me are enclosures. The only one with any very distinct connection with history or legend contained a dwelling with apartments ... the glazing or cementing is done systematically, and instead of using mortar.[37]

An account of the buildings is provided in Angus Smith's book, *Loch Etive and the Sons of Uisnach*, which also includes a letter from Joseph Black on the nature of the vitrification of these forts.[38] From analyses carried out by Frank Scudder, Angus Smith's assistant, the glaze contained silica (68.88 per cent), alumina (16.17 per cent) and iron (5.33 per cent). Angus Smith drew attention to some 50 sites in Scotland, two or three in Ireland, at least one in Bohemia, some in France, perhaps even some in the Euphrates valley and other parts of Asia, although the exact locations are not identified.

Whenever pressure from his professional duties allowed, Angus Smith would readily and enthusiastically pursue his antiquarian interests. In 1872 he joined James Young and his yacht *Nyanza* on a voyage to Iceland, and details of the trip were recorded in the book *To Iceland in a Yacht*; a similar voyage to St Kilda in 1872 resulted in the book *Visit to St. Kilda in the 'The Nyanza'*.[39] Angus Smith steadily built up a reputation as a respected antiquarian with a fine library and in 1874 was elected a Fellow of the Society of Antiquaries of Scotland.[40] Besides his contribution to the Society of Antiquarians of Scotland, he was elected one of the two Vice-Presidents of the Lancashire and Cheshire Antiquarian Society at its formation in 1884, but there is no evidence that he contributed to its proceedings.

---

[37] Robert Angus Smith, 'On a vitrified mass of stone from the Fort of Glen Nevis', *Proceedings of the MLPS*, 22 (1883), pp. 1–3.

[38] Angus Smith, *Loch Etive*, p. 367.

[39] Robert Angus Smith, *To Iceland in a Yacht* (Edinburgh, 1873), including a series of delightful drawings by his niece, Jessie Knox Smith, and Robert Angus Smith, *Visit to St. Kilda in the 'The Nyanza'* (Glasgow, 1879).

[40] Anon, 'Obituary: Robert Angus Smith', *Proceedings of the Society of Antiquaries of Scotland* XIX (1884–85) p. 6.

Angus Smith's wide interests are also reflected in his extensive library of 3,946 books.[41] Besides the expected collection of books on chemistry, metallurgy, medicine and sanitation, the library covered subjects such as philology (346 volumes), history (721), with 84 volumes on Celtic history, natural history (39) and poetry (125). There were 47 guidebooks and many books in foreign languages that reflect his interest in travelling through Britain and mainland Europe and keeping abreast of events generally in other countries. There were many novels and plays (114) and, intriguingly, 83 books relating to the Italian Statistical Society. Interestingly for an experimental scientist, 89 books relate to occult science, confirming Angus Smith's dalliance with spiritualism. Further details of the library are provided in Chapter 8.

From the mid-nineteenth century society showed a growing fascination with spiritualism, and the shift from the material to the spiritual world brought an interest in such psychic phenomena as communicating with the spirits of the dead, table turning and telepathic action. Such phenomena caught the attention not only of the public but also many scientists from a range of different disciplines. These psychic phenomena were linked in the scientific mind to other controversial areas of scientific investigations such as spectra, electrostatic electricity and telegraphic communications, areas that remained the focus of intense speculation and debate through the rest of the nineteenth century.

Prominent scientists reacted in markedly different ways to the spiritualism phenomena after experiencing and reviewing some of the alleged psychic occurrences. Thomas Huxley as a biologist and agnostic felt that spiritualism offered nothing more than amusement with no serious scientific basis; Michael Faraday as a natural philosopher and Sandemanian abhorred spiritualism as an offence to religion and to science.[42] However, as Richard Noakes has suggested, many scientists 'sought to raise their social and economic status by exploiting their laboratory control over unstable and controversial phenomena'.[43] These scientists included Cromwell Varley (electrician), William Barrett (physicist) and William Crookes (spectroscopist). Crookes is of particular interest because his involvement in spiritualism is most probably linked to Angus Smith.

---

[41]   After Smith's death the library was purchased by friends and donated to Owens College. The collection later transferred to the University of Manchester Institute of Science and Technology (UMIST) and more recently to The John Rylands Library, University of Manchester. A full list of the books in the library has not survived, but it is possible to identify some of the individual books through the library's electronic and card catalogues, and by a bookplate that reads: 'Manchester Owens College / Presented by / The Angus Smith Memorial Committee / 1884'.

[42]   Jack Meadows, *The Victorian Scientist: The Growth of a Profession* (London, 2004), pp. 178–80.

[43]   Richard Noakes, *'Cranks and Visionaries': Science, Spiritualism and Transgression*, PhD diss., University of Cambridge, 1998, p. 4.

Figure 1.2   William Crookes (1832–1919), chemist and science journalist, in
the 1870s when investigating psychic phenomena
Courtesy of the Royal Institution of Great Britain

It is not known with certainty when Angus Smith became interested in psychic
phenomena, nor who or what had prompted his interest. Nevertheless, it appears
that Crookes was persuaded by Angus Smith following the death of his younger
brother, Philip Crookes, from yellow fever during the laying of a telegraph cable
between Cuba and Florida. This is inferred from a letter dated 22 December 1869
from Crookes to John Tyndall that refers to his friend who is a Fellow of the Royal
Society and stands in the 'foremost rank of experimental scientists'.[44] This letter

[44]   Crookes to John Tyndall, 22 December 1869. See Tyndall's Journal (Royal
Institution), f. 1364, and Richard G. Medhurst, Kathleen M. Goldney and Mary R.
Barrington, *Crookes and the Spirit World* (London, 1972), pp. 232–4. Crookes refers to the

may refer to the first known communication on spiritualism between Angus Smith and Crookes, who were regular correspondents, a letter dated 9 April 1869, and Bill Brock has suggested that it was Angus Smith who recommended that Crookes read Epes Sergent's book *Planchette*, a book that was to greatly influence Crookes and his belief in psychic phenomena.[45]

Crookes was determined to subject the psychic phenomena to scientific investigation, but this required the cooperation of the mediums and mystics in an experimental environment, and after several unsuccessful attempts he managed to persuade the medium Daniel Home to come to his house in Mornington Road, London for dinner followed by a séance to which Angus Smith and Alfred Russel Wallace, the naturalist and evolutionary theorist, were invited. It is not known what Angus Smith's reaction was to the séance, but Crookes, who scrutinized the séance closely and used instruments to test for the existence of any electrical phenomena, seemed convinced of the special powers of Home and perhaps over-optimistically believed that he would be able to persuade scientific colleagues of its substantiation.[46] However, many years later, when papers reporting these experiences were presented at the Bristol meeting of the British Association for the Advancement of Science in 1898, there was general astonishment and incredulity that such subjects were allocated time in the meeting programme, even though scientists such as Wallace, Barrett and Oliver Lodge had lent their support to the airing of these controversial subjects. Several scientists, most notably Crookes, soon came to realize that their professional scientific standing was being undermined by their interest in spiritualism. This is probably the main reason why Angus Smith never wrote widely about spiritualism, or indeed considered writing a book as Crookes did.[47] With science underpinning his work as a government inspector, Angus Smith quickly realized that it was very important not to alienate those who relied on his professional advice, such as fellow scientists, manufacturers and businessmen, civil servants and parliamentarians. Nevertheless, Angus Smith's collection of 89 books on the occult is testament to his enduring and private fascination with spiritualism. He was a member of the Society for Psychical Research from 1882 to 1884.

---

person who introduced him to spiritualism 'some six months ago' as an 'FRS ... foremost in the rank of experimental philosophy'.

[45]   Crookes reported the influence of this book to its author. Crookes and Angus Smith were in regular contact by letter and at BAAS Annual Meetings. In 1864 Crookes had written to Angus Smith to explore the opportunities in Manchester for a consulting chemist, to which Angus Smith replied; see E.E. Fournier-d'Albe, *Life of Sir William Crookes* (London, 1923), pp. 86–7 and 105.

[46]   Noakes, *'Cranks and Visionaries'*, pp. 186–7. For a useful summary of Crookes's writings on psychic phenomena, see 'Sir William Crookes on Psychic Research', *Smithsonian Report*, 1898, pp. 185–205.

[47]   Crookes destroyed most of his correspondence on spiritualism. From Crookes's letter book it is known that Angus Smith wrote him 15 letters on the subject.

Angus Smith remained loyal to Manchester, retaining his home and separate laboratory there. He never married, but his niece Jesse Knox Smith (daughter of his brother John) acted as housekeeper for quite a long period of his life. He was a gentle and caring person:

> His temper was singularly even and placid: he had his checks and crosses, of course, like other men, and he was occasionally pained to find himself misunderstood. But nothing ruffled his calm. His perfect transparency, his charming simplicity, and a certain quiet playfulness of manner gained for him the sobriquet of 'Agnus' Smith. Indeed, his sense of fun could see the latent humour in any situation.[48]

His close friend, Edward Schunck drew attention to Angus Smith's moral character:

> in his case an intellect of high order was united to a character of the purest and noblest type. The most marked trait in his character, it always seemed to me, was a wide benevolence, a benevolence which seemed capable of embracing all except the unworthy within its fold ... His extreme conscientiousness and high sense of honour appear even in his works, leading him scrupulously to weigh all that could be said on either side of the argument, and to give every man his proper share of merit, refusing sometimes even to credit himself with what was manifestly his due.[49]

Most of his social activities revolved round meetings of the MLPS and its specialist sections such as the Microscopical and Natural History section. He satisfied his intense interest in Scottish history and archaeology by making regular visits north of the border whenever time allowed from his busy schedule, and these visits were often combined with the opportunity to visit friends or relatives. But one of his greatest joys was sharing adventurous pursuits with his lifelong friend James Young, whose luxury yacht more often than not played a part in the adventures.[50]

Besides his PhD and FRS, Angus Smith's honours included LLD degrees from Glasgow University and from Edinburgh University, Corresponding Member of the Royal Bavarian Academy and the Imperial Geological Institute of Vienna, and a member of the Juries for the London Exhibition for 1862, and for the Exposition in Paris, 1878.[51]

As mentioned earlier, Angus Smith's health had been in decline for some time, but in January 1884 he became completely exhausted from the exertions of his

---

[48]   T.E. Thorpe, 'Robert Angus Smith', *Nature*, 30 (1884), p. 105.

[49]   Schunck, 'Angus Smith', pp. 100–101. It was Edward Schunck who arranged for a bronze bust to be sculpted by Thomas Nelson McLean (1845–1894) and presented to the MLPS.

[50]   Smith, *To Iceland in a Yacht*.

[51]   'Dr. Angus Smith, FRS', *The Biograph and Review*, 5 and 6 (1879–81), p. 152, and 'Dr. Angus Smith', *The Manchester Guardian*, 13 May 1884.

professional work, principally from the demands of his two government inspector posts. Nevertheless, while recuperating in Matlock Bath, Derbyshire, he wrote to the Secretary of the Local Government Board about his determination to retain his inspector posts:

> after only a month's holiday in Norway I struggle onwards determined ... I am so far down that I am carried to my bedroom and sitting room, a condition hitherto unknown to me. Still my head is clear and my investigations have come to a point where I am excited to do more. I am very unwilling to lose hold of the reins either in the Alkali or Rivers Pollution Prevention Departments.[52]

While never giving any intimation of wanting to retire, Angus Smith was granted a period of leave and, although some responsibilities were delegated to the sub-inspectors within the Alkali Inspectorate, Angus Smith was adamant about being consulted on important policy matters:

> Any difficult question may be sent to me or be brought to me if I go home, but this arrangement will enable me to attend or otherwise to important questions which may cause too much anxiety as I am of an anxious disposition at any rate.[53]

Later in March he went to stay at Glynwood in Colwyn Bay (north Wales) for a further period of recuperation, but died there on 12 May 1884 of complications brought on by pernicious anaemia.[54]

Angus Smith's funeral took place on 16 May, with many of those attending representing the various organizations with which Angus Smith was associated at different stages of his life.[55] Besides family relatives, including Jessie Knox Smith (niece and long-time housekeeper), the Alkali Inspectorate was represented by several of the sub-inspectors with whom Angus Smith had worked so closely – Alfred Fletcher, Charles Blatherwick and George Davis. The Society of Chemical Industry was represented by its President, Ivan Levinstein. Others attending included Henry Roscoe (representing the MLPS), Alexander McDougall, relatives of the Spence family, Frederic Scudder (Angus Smith's personal assistant), relatives of James Young, James Fenwick Allen and Mrs Schunck representing her husband, who was unable to attend. The funeral procession made its way from Angus Smith's home, Gowrie House in Oxford Road, to St Pauls's Church in Kersal, where he was buried with a simple headstone.[56] Robert Angus Smith's legacy is discussed in Chapter 8.

---

[52]   Letter from Robert Angus Smith at Matlock Bath, Derbyshire to the Secretary of the Local Government Board, dated 13 January 1884. NA: (Ref: MH16/2).

[53]   Ibid.

[54]   His personal estate was valued at £1,953 15s 4d (£1953 77p).

[55]   'The Late Dr. Angus Smith', *The Manchester Guardian*, 17 May 1884.

[56]   The headstone marking Robert Angus Smith's grave remains to the present day.

Emergence of the Civil Scientist

The first half of the nineteenth century brought unprecedented changes to British society through the continued process of mass industrialization from the previous century.[57] Although a small section of society reaped major benefits from these changes, the majority were left facing great hardships, embracing health, housing, provision of food, as well as exposure to dangers inherent in the profusion of new steam-powered technology. Society was constantly made aware of these hardships through day-to-day experiences and through the writings of authors such as Charles Dickens, Henry Mayhew, George Eliot and Friedrich Engels. Others, of a Benthamite leaning, formed a close-knit network of like-minded reformers who were supportive of each other's actions as part of a long-term programme of improvements. The state was forced to respond through parliamentary legislation in an attempt to embrace the initiatives and ideas of the reformers in trying to alleviate some of the worst of the shocking evils that formed part of everyday life for much of the population.

As historian Oliver McDonagh has shown, the legislative response during the first half of the nineteenth century evolved into a standard model.[58] A crucial element of this model was the incorporation of some form of regulation into the parliamentary legislation so that its terms were enforceable. However, strict enforcement required 'executive officers', or inspectors, as they became more generally known, to monitor compliance with such regulations, and, in the event of non-compliance, they could seek financial penalties through the judicial system. Increasingly, legislation targeted different sectors of society, including factories, railways, mines, health and later various branches of manufacturing industry. Such a broad canvas required inspectors with appropriate knowledge and experience to carry out their duties effectively; they were subsequently to draw on a broad range of experts and professional people, which would lead to some inspectors with the designation 'civil scientist'.

---

[57] I have avoided referring to the 'industrial revolution'. Although the phrase provides a useful peg on which to hang ideas about mass industrialization from the eighteenth century, there is increasing scepticism among historians that an 'industrial revolution' took place between particular dates, e.g. 1750–1850, and may in fact have occurred in a more evolutionary manner over a much longer period of time. See David Cannadine, 'Engineering History, or the History of Engineering? Re-Writing the Technological Past', *Transactions of the Newcomen Society*, 74 (2004), pp. 167–70.

[58] Oliver McDonagh, 'The Nineteenth Century Revolution in Government: A Reappraisal', *The Historical Journal*, 1 (1958), pp. 52–67.

Changes in Society and Transformation of the State within Society

Although mass industrialization had its origins in the middle of the eighteenth century, its impact was progressive; and as the process of industrialization accelerated and intensified, so the effects on society deepened. As McDonagh has pointed out, a transformation was brought about by

> the vast increase, and new concentrations and mobility, of population ... the developments in mass production and cheap and rapid transport, by the new possibilities of assembling great bodies of labour, skills and capital, and by the progress of the technical and scientific discovery associated with this economic growth.[59]

These dramatic and far-reaching changes to society were accompanied by exposure to dangers, harsh and unhealthy living conditions, and practices felt to be against the humanitarian spirit. The state gradually found that it had no choice but to engage with the growing problems. The government had to take action to ameliorate the worst excesses that were being wrought on society. This required Parliament and politicians to be more sensitive to society's needs and take appropriate action using investigative instruments and legislative powers. It is against this backdrop that some of the most far-reaching reforms and innovative legislation in Britain were enacted. The process was not confined to just the first 50 or so years but continued as Parliament felt necessary and desirable right up to the present day. The early influential legislation included the Factories Act (1833), Reform Acts (1832, 1867 and 1884), Repeal of the Corn Laws (1846), Public Health Act (1848) and Education Act (1870). In its own way, the Alkali Act was also important in providing a check on the way chemical processes (initially at alkali works) were operated and brought about strict control of the noxious gases and vapours released into the atmosphere that caused damage to surrounding property. Significantly, it was only later that consideration was given to any harmful effects on human health. Reforms often relied on an individual reformer or campaigner who would lead the way and agitate to ensure that the necessary changes and improvements were implemented for the benefit of society as a whole. Leaders in the campaigns included Lord Shaftesbury, Edwin Chadwick, John Simon and Richard Cobden. This was truly an Age of Reform.

Emerging Inspectors

With the changing role of the state in addressing some of the harmful effects of industrialization over the first half of the nineteenth century, inspectors were appointed for schools of anatomy (1832), factories (1833), lunacy (1833),

---

[59]   Ibid., p. 57.

emigration (1833), Poor Law (1834), prisons (1836), tithe commutation (1836), education (1839), railways (1840), mines (1842), public health/local government (1848) and mercantile marine (1850). As historian of government inspection Peter Bartrip has pointed out:

> a major problem with discussing inspection arises from the difficulties of defining what is meant by the term, for inspectorates differed in terms of purposes, powers, duties, organization, status, qualification, titles, jurisdictions and remuneration.[60]

Inspectors were essentially enforcement officers, and this role brought difficulties during the appointment of inspectors. While it was important to have well-qualified inspectors with experience of the field of inspection, it was crucial that they did not have any financial interest or close associates in the field that might undermine their authority. Mine inspectors were usually drawn from those who had mine-working experience and railway inspectors usually came from the Royal Engineers because of their prior experience with general engineering and bridge construction.

It is important to emphasize that the inspectors appointed under the Alkali Act 1863, who as a group became known as the Alkali Inspectorate working within the Board of Trade, were the first scientists to be appointed as inspectors. While most other groups of inspectors safeguarded people or groups of people, the Alkali Inspectorate was required to safeguard property alone. However, towards the end of the nineteenth century the alkali inspectors joined forces with factory inspectors and public health officers to ensure greater scrutiny of the damage done to health in the working and living environment brought about by a range of noxious gases from chemical processes.

Professionalization of Science and Scientists

An important transformation in society during the nineteenth century, with tremendous ramifications for the role of inspectors, was the professionalization of science and scientists. It is interesting to note that it was only in 1834 that the Cambridge polymath William Whewell proposed adoption of the word 'scientist' to embrace all those engaged in scientific activities. However, many refused to use the word to describe their work, preferring 'natural philosopher' or 'man of science'.[61] Professionalization was achieved in a number of interconnected ways; these included appearing as expert witnesses in court cases or parliamentary

---

[60]   P.W.J. Bartrip, 'British Government Inspection, 1832–1875: Some Observations', *The Historical Journal*, 25 (1982), p. 606.

[61]   William Whewell, Review of 'Art, III – On the Connexion of the Physical Sciences, by Mary Somerville', *Quarterly Review*, 51 (March 1834), pp. 59–60. Michael Faraday

inquiries; providing scientific advice to major bodies; carrying out scientific consultancy work; expanding the role of science within universities; and founding of scientific institutions either to further science generally or to promote the interests of individual sciences. Together they brought about a major change in the standing of both science and scientists during the second half of the nineteenth century and were contributory factors in the creation of the civil scientist.

From the early part of the nineteenth century those engaged in science were often called to give evidence in court cases, whether for the prosecution or the defence, in which their evidence might help to resolve some legal claim. In April 1820, at the court case of *Severn, King and Company (sugar bakers) versus the Imperial Insurance Company* to recover losses arising from a major fire, the expert witnesses included Samuel Parkes (chemical manufacturer), William Brande (Professor of Chemistry at the Royal Institution) and Friedrich Accum for the plaintiffs; and Richard Phillips (Professor of Chemistry at the Royal Military College) and Michael Faraday (Royal Institution) for the insurance company.[62] Other high-profile cases involving chemical evidence took place in 1838 when Thomas Thomson, Professor of Chemistry at Glasgow University, appeared in a nuisance court case giving evidence for James Muspratt, the Liverpool chemical manufacturer, while in 1857 Angus Smith and Edward Frankland appeared for the defence and prosecution respectively in the *Regina v Spence* court case. Too often in these cases the judge and jury were left floundering and unable to differentiate between the conflicting evidence from scientists. Nevertheless, gradually these cases demonstrated the value of scientific evidence and thereby added to the standing of the scientific person providing such evidence. These cases also involved the medical profession, particularly in cases of possible malpractice.

The reputation of scientists was also enhanced by their involvement in parliamentary inquiries, as a few examples will illustrate. In 1843 both Lyon Playfair and Angus Smith contributed to the report on the Large Towns of Lancashire for the Health of Towns Commission; in 1843 Michael Faraday gave evidence on behalf of the Ordnance Office into the cause of an explosion at Waltham Abbey gunpowder factory; in 1848 Angus Smith completed research for the Royal Commission on the Sanitation of the Metropolis; in 1864 he completed research for the Royal Commission on Mines; in 1865 both he and William Crookes carried out research for the Royal Commission on Cattle Plague that reported in 1865 and 1866.

For those scientists without private financial means, which in fact was the majority, consultancy work was important. Between 1836 and 1865 Michael Faraday acted as scientific adviser to the Corporation of Trinity House, providing his scientific knowledge and experience to aid the development of lighthouses

---

preferred to be known as a 'natural philosopher', while John Tyndall preferred 'man of science'. Many of those engaged in physics were uncomfortable with 'physicist'.

[62]   June Z. Fullmer, 'Technology, Chemistry, and the Law in the Early 19th-Century England', *Technology and Culture*, 21/1 (1980), pp. 1–28.

around the coasts of England, Wales and the Channel Islands.[63] In 1844, when Lyon Playfair left Manchester to take up an appointment at the Geological Survey in London, Angus Smith decided to remain in Manchester and pursue a consultancy business with a wide range of services that included carrying out routine chemical analysis for a variety of clients, advising on the operation of chemical works, investigating air quality, analysing water for railway companies and sanitary authorities, and offering advice generally on scientific matters. With Manchester developing as a major industrial metropolis, it was to prove a very good base for Angus Smith. As his name became more widely known, so his consultancy work extended well beyond the city boundaries to give him a national presence.

The standing of science and scientists was also enhanced by the expansion in university education during the nineteenth century. For chemistry, an important event occurred in 1845 with the founding of the Royal College of Chemistry in London. It had been possible to study chemistry earlier at mechanics institutes, Scottish universities, medical colleges, University College London (1826), King's College London (1828), University of Durham (1833) and also at Oxford and Cambridge, although in the main the subject was studied as part of a medical qualification. The Royal College of Chemistry adopted the approach taken by Justus von Liebig at the University of Giessen that combined formal instruction with practical exercises, applying the knowledge so gained to genuine and original practical research projects. It was this approach that had attracted so many from different countries to Giessen. After graduating from Giessen, these students took up influential positions in industry, consultancy work or teaching. This was the course taken by Angus Smith, who attended Giessen in 1841 and 1842.

Instruction at the Royal College of Chemistry was directed by the German-born and -trained A. Wilhelm Hofmann, who had been a fellow student of Angus Smith at Giessen. It was while attending the London College that, in 1856, William Perkin discovered the first synthetic coal-tar dye, aniline purple, later known as mauve. This discovery led to the establishment of Britain's coal-tar dyestuffs industry, although from the 1860s it fell behind Germany's. Other chemists trained at the College went into the British chemical industry, which had previously often relied on chemists trained in Germany or other countries. Royal Manchester Institution began its chemistry courses under Lyon Playfair's direction in 1843 with Angus Smith as his assistant. By 1877, the year of the foundation of the Institute of Chemistry, some 60 educational institutions were providing chemistry courses in Britain.[64]

With more people gaining qualifications in science, the nineteenth century also saw the foundation of a number of scientific institutions that promoted the benefits of science and enhanced the professional standing of scientists. In 1831 the British Association for the Advancement of Science (BAAS) was founded in York and

---

[63]   Scotland and Ireland had their own authorities.

[64]   Colin A. Russell, with Noel G. Coley and Gerrylynn K. Roberts, *Chemists by Profession* (Milton Keynes, 1977), pp. 89–91.

held an annual meeting in a different city each year, thus bringing discussion of the latest ideas in science to the wider population of Britain rather than just focusing on London or a few major centres.[65] The BAAS divided its activities into a number of discrete sections, each one covering a specific body of science or group of sciences. Right from the very first meetings a prominent scientist in a particular field was appointed president of his section for the year of the meeting. Michael Faraday was President of the chemical section in 1837 and 1846, and Vice-President of the Association in 1844, 1849 and 1853.

Several specialist scientific societies were formed in London during the nineteenth century. These included: Linnean Society (1788), Geological Society (1807), Astronomical Society of London (1820) (but becoming the Royal Astronomical Society in 1831), Zoological Society (1826), Botanical Society (1836) and the Chemical Society (1841). Later in the century other chemical societies were founded to reflect further specialization, including the Institute of Chemistry (1877) for those active in analytical chemistry and the Society of Chemical Industry (1881) for those working in the chemical industry.[66] Although based in London, these societies had regional sections providing a range of professional and social activities in furthering their objectives. Angus Smith served on the council of the Chemical Society in 1870–72 and was Vice-President in 1878–80; he was also Vice-President of the Institute of Chemistry. He was a member of the committee appointed on 19 April 1880 to pursue 'a society for the promotion of the application of chemical science to manufacture', which led to the formation of the Society of Chemical Industry in 1881. Angus Smith was to take a prominent part in the activities of the Manchester section of the new Society.[67]

With parliamentary legislation increasingly including regulations with a strong scientific or technical element to enable their enforcement, and scientists demonstrating the important part science could play in the affairs of society, Parliament and government embraced an increasing role for scientists to act as inspectors. Evidence given during the Select Committee on Noxious Vapours in 1862 had clearly demonstrated the important part science and scientists could play in regulating industrial processes. So, when the Alkali Act gained the Royal Assent in 1863, the appointment of a scientist to act as Inspector did not raise any

---

[65]   It is only recently that the British Science Association (in 2009 the BAAS was rebranded as the British Science Association) decided to rotate the annual meeting among just five major cities serving the regions of Britain.

[66]   In 1972 the Chemical Society and the Royal Institute of Chemistry amalgamated to form the new Chemical Society, and in 1980 the Royal Society of Chemistry was formed by the merger of the Chemical Society, the Royal Institute of Chemistry, the Faraday Society and the Society for Analytical Chemistry with a new Royal Charter and the dual role of learned society and professional body. The Society of Chemical Industry remains a separate organization.

[67]   D.W. Broad, *Centennial History of the Liverpool Section Society of Chemical Industry 1881–1981* (London, 1981), p. xxiv.

concerns. When in February 1864 Robert Angus Smith was appointed Inspector under the 1863 Alkali Act, he became one of the first scientists appointed to this civil scientist role with its overriding responsibility to enforce the terms of parliamentary legislation rather than just act in an advisory capacity.[68] While understanding the nature and origin of the civil scientist designation forms an important part of this introduction, reading the remainder of this book will give the reader a more thorough assessment of Robert Angus Smith's success in this important role.

---

[68] Many other scientists might be considered civil scientists, such as the Astronomer Royal, but not in the sense defined here, to enforce the terms of parliamentary legislation.

Chapter 2

The Influence of Graham, Liebig and Playfair on Robert Angus Smith

From the late 1820s, Thomas Graham, the German chemist Justus Liebig and Lyon Playfair were to exert a major influence on British science, both in advancing understanding and increasing the role of education, and in the application of science for the wider benefit of society. This was at a time when new chemical ideas emerging from the late eighteenth century were beginning to bring major economic benefits. Each man played his own individual role but also acted most effectively in concert, as with work on behalf of the British Association for the Advancement of Science (BAAS from now on) 1831. Graham, Liebig and Playfair were to influence many British scientists in the first half of the nineteenth century, and none more so than Angus Smith in many different ways: Graham's lectures at Anderson's University; Liebig's chemistry course at Giessen and his promotion of science for the benefit of society as a whole; Playfair's work in Manchester with the Royal Manchester Institution (RMI from now on), the Royal Commission on the Health of Towns and the Manchester Literary and Philosophical Society (MLPS from now on). For Angus Smith, their influence came at key times in his life, often when guidance and advice were to prove crucial.

Thomas Graham, Glasgow and London

Thomas Graham was born in Glasgow in 1805, the son of a prominent and prosperous merchant.[1] When Graham attended Glasgow University in 1818 to study the arts, he came under the influence of Thomas Thomson, then Professor of Chemistry at the University, who fired Graham's interest in chemistry.[2] Later Thomson convinced the university authorities of the benefits of a laboratory for teaching chemistry and this led Thomson to become 'the first teacher of

---

[1] For details of Graham's life and work, see Michael Stanley, *The Chemical Work of Thomas Graham*, PhD diss., Open University, 1980; Michael Stanley, 'Entry for Thomas Graham', *ODNB*; and Robert Angus Smith, *The Life and Works of Thomas Graham* (Glasgow, 1884).

[2] Thomson was one of the outstanding chemists in the 1830s and appeared as an expert witness for James Muspratt in his 'nuisance' court case in Liverpool in 1838.

Figure 2.1   Thomas Graham (1805–1869), chemist and Master of the Mint, photograph by Maull and Polyblank
The British Library Board

practical chemistry in a British university' when the laboratory opened in 1818.[3] Graham knew of these developments, being by this time Professor of Chemistry

[3]   Jack Morrell, 'Entry for Thomas Thomson', *ODNB*. See also Jack Morrell, 'The Chemist Breeders: The Research Schools of Liebig and Thomas Thomson', *Ambix,*

at Anderson's University in Glasgow, and this laboratory-based teaching of chemistry was to greatly influence Graham's own approach.[4]

Graham's father wanted his son to become a Church of Scotland minister but Thomas resisted wherever possible and retained as best he could his commitment to chemistry. With subterfuge in mind, in 1826 Graham persuaded his father that divinity was better taught at Edinburgh than at Glasgow, while really intending to study medicine (then the best route into chemistry) and spend as much time as possible working in the chemical laboratory there. Later his father became so impatient with his son's tardy commitment to become a minister that he stopped his financial support and Graham was forced to teach mathematics. In 1831 Graham read a paper on diffusion of gases to the Royal Society of Edinburgh, and invited his father to the Edinburgh lecture. His father, sensing in the formal surroundings what his son had achieved, withdrew his objections and they were reconciled. This conflict between a parent's expectations of his son entering the Church and the son's wish to follow another career (in Graham's case, chemistry) was also experienced by Angus Smith, although in his case the turbulence was intensified and persisted over a longer period, and perhaps never finally resolved in his mind.

Graham's lectures at Anderson's University proved very popular and influential in inspiring many students to pursue careers in chemistry. The lectures generated a good deal of excitement and general interest among the public. A young Robert Angus Smith, then aged about 12, was enticed to attend a few of the lectures, possibly with encouragement from his brother Joseph with whom he shared an interest in chemistry:

> I was told by a boy ... that a wonderful young man had appeared, one who, although only nineteen years old, had still a reputation. I went to hear one or two of his popular lectures, and I remember his looks well. He might have been taken for nineteen, but in reality, he was twenty-five when he first lectured at the Andersonian. He had the same quiet, rather stiff, and hesitating manner which he never lost. He did not cause enthusiasm by brilliance of address, but a certain reserve and a certain feeling of power, as well as of ambition, which his letters prove to have been strong in him, so acted on his demeanour, that students became attracted, and were ready to work beside him and devote themselves also to his service.[5]

---

19 (1972), pp. 1–46, and J.B. Morrell, 'Thomas Thomson: Professor of Chemistry and University Reformer', *British Journal for the History of Science*, 4 (1969). pp 245–65.

[4]   Anderson's Institution was renamed Anderson's University in 1828, and in 1964 became the University of Strathclyde.

[5]   Angus Smith, '*Thomas Graham*', p. 65. On 20 March 1884 Angus Smith was due to deliver the Second Graham Lecture to the Philosophical Society of Glasgow – Chemistry Section at University of Glasgow, but because of ill health he was unable to attend and so the lecture was read for him.

Many students besides Angus Smith were to become prominent chemists in either academia or industry after attending Graham's lectures at Anderson's University. These included James Young, Lyon Playfair, John Stenhouse, Walter Crum and the Muspratt brothers, Richard and Sheridan.[6] David Livingstone, the future Africa explorer who was studying medicine at Anderson's University in 1836, probably attended some of Graham's lectures at the insistence of his friends, James Young and Lyon Playfair.[7] Young, who had become Graham's assistant in the 1831–32 session, may have been known to Angus Smith since it has been suggested that they may have been childhood friends as they lived near each other. True or not, they became lifelong friends, sharing an interest in scientific matters, Celtic history and especially travel. Later, through Playfair's intervention, Young became a leading figure in the extraction of oil from coal-shale that made him very wealthy and resulted in his sobriquet, 'Paraffin Young'.

In 1837 Graham was appointed Professor of Chemistry at University College London, the same year that the annual meeting of the BAAS was held in Liverpool. Annual meetings of the BAAS brought together many of the well-known practitioners of science with their local counterparts. One of the keynote speakers at the Liverpool meeting was the German chemist, Justus Liebig, whose pioneering research advanced the study of organic chemistry and who had established a chemistry course at the University of Giessen that involved students developing their chemical knowledge and understanding through a series of carefully set-out practical laboratory experiments, followed by a research project that brought new understanding and knowledge. The first part had similar features to Thomas Graham's course in Glasgow and then in London, but it is generally accepted that the courses were developed independently.

By the time of the Liverpool meeting Liebig was moving away from pure chemistry and was concentrating on applications of chemistry, especially to agriculture and physiology, to benefit society more widely. This was the subject of his exciting address to Section B, but besides the formal proceedings there were opportunities for informal conversations and visits. Liebig met Graham on this occasion and they both expressed admiration for the other's work that would become the basis for long-term respect and support. During the planned visits, Liebig visited the extensive alkali works of James Muspratt in Vauxhall Road, in the centre of town. Liebig was so impressed by the scale and organization of

---

John Stenhouse, FRS, studied at Giessen before becoming a lecturer in chemistry at St Bartholomew's Hospital and later assayer to The Royal Mint; Walter Crum, FRS, was a businessman in calico printing and used Turkey red dyeing; Richard Muspratt was an alkali manufacturer in Flint, North Wales; Sheridan Muspratt studied at Giessen before founding the Liverpool College of Chemistry in 1848.

7   It has been suggested that Angus Smith knew Lyon Playfair and David Livingstone through attending Graham's lectures at Anderson's University but this is unlikely because Angus Smith was in London living with his brother at this time. Playfair remembered Young and Livingstone but not Angus Smith. See Reid, *Lyon Playfair*, p. 36.

the works that a few years later he and James Muspratt were to collaborate on manufacturing artificial manures based on Liebig's work on investigating minerals found in various plants. Unfortunately, the manures failed commercially because their physical consistency prevented the minerals from leaching effectively into the soil. Advances in the efficacy of artificial manures relied more on the innovations by John Bennet Lawes's work at Rothamsted in the early 1840s; this is discussed in Chapter 7.

In 1837, when Graham left Glasgow for London to take up his appointment at University College, one of his outstanding young students, Lyon Playfair, decided to move to the University of Edinburgh to continue his medical studies and his interest in chemistry. Unfortunately, having suffered severe eczema from working in the dissection room, he was unable to complete his medical course. Playfair was persuaded by his father, Chief Inspector of Hospitals in Bengal, to return to India, where he had been born, to take up a merchant's clerk post. However, unimpressed by the nature of the work and its prospects, Playfair returned to England in 1838 and searched in London for Graham, who was by this time beginning to make a name for himself there. By this time also, Graham had appointed James Young (another of his bright Glasgow students) as his assistant. But not wishing to lose the services of Playfair, who had received chemistry prizes in Graham's classes in Glasgow, Graham asked him if he would be his personal laboratory assistant, and Playfair readily accepted. The following year, realizing that Playfair would benefit from broadening his chemical experience, Graham suggested that he went to study with Liebig in Giessen and in October 1839 Playfair registered as a student there. This was during a period when most of the Giessen students were working on fats of one kind or another following Chevreul's influential work. While in Giessen, Playfair discovered a new fatty acid in the butter of nutmeg and a new substance in cloves; he matriculated with a PhD in February 1841.[8] But, as mentioned earlier, besides the benefits of working with Liebig and obtaining his PhD, there were the opportunities to meet other aspiring young chemists such as Edward Schunck and Graham's former students, John Stenhouse and Robert Angus Smith.

Justus von Liebig, Giessen and the British

Justus von Liebig was born in Darmstadt on 12 May 1803.[9] His initial education was in Germany but between 1823 and 1825 he continued his studies in Paris with Charles Bernard Desormes, Pierre Louis Dulong, Joseph Louis Gay-Lussac and Louis Jacques Thenard, and worked in the laboratory of the Ecole Polytechnique.

---

[8]   The new fatty acid in butter of nutmeg was named 'myristic acid' and the crystalline substance in cloves was named 'caryophylline'. For details of Playfair's time studying with Liebig at Giessen, see Reid, *Lyon Playfair*, pp. 42–3.

[9]   Liebig was ennobled in 1845.

Figure 2.2   Justus von Liebig (1803–1873), chemist, oil painting by Wilhelm
               Trautschold
               Wellcome Library, London

These opportunities provided Liebig with the dual benefit of a good grounding in some of the latest ideas in chemistry and useful practical laboratory experience.[10] He received a PhD from Erlangen for this research on 'The Relation of Mineral Chemistry to Plant Chemistry' in 1823, and the following year he was appointed Associate Professor of Chemistry at the University of Giessen, which began an association with Giessen until his move to Munich in 1852.

At Giessen, Liebig established his laboratory and the internationally renowned school of chemistry that attracted many aspiring young chemists from across the world. The unique training Liebig provided had two strands: developing understanding and knowledge of chemistry through a series of carefully structured practical laboratory exercises monitored by Liebig's assistants, followed by undertaking a research project defined and mentored by Liebig that allowed the knowledge and understanding to be applied to make new discoveries. It was a major change from the lecture-centred approach then practised in most countries, and it was a model that many of Liebig's students were to pursue in their own institutions when they returned home.[11] As mentioned earlier, it was not just the opportunity to work with Liebig and his team of assistants, but the camaraderie between students drawn from all over the world that so often remained as lifelong friendships, in some cases passing on within families to the next generation.[12] A bond was created between students that remained steadfast and resilient when opportunities subsequently arose for them to work together. Attendance at Giessen provided benefits that had a profound influence on British science, as Thomas Graham claimed in a letter to Liebig in 1854:

> More students from this country than from any other land beyond the bounds of Germany, have worked in the laboratory of Giessen, and have derived incalculable benefit from the institution there imparted, and from the noble example there presented to them of an elevated philosophical and scientific life.[13]

Liebig had developed a simple method of quantitative analysis of organic compounds that opened up a better understanding of the natural world. He was a pioneer in the application of chemistry to physiology and agriculture. With the latter he examined the mineral nutrients in the soil and in plants and crops with

---

[10]   Walter Killy and Rudolf Vierhaus (eds), *Dictionary of German Biography* (Munich, 2003), pp. 399–400. See also Ralph E. Oesper, 'Justus von Liebig – Student and Teacher', *Journal of Chemical Education*, 4/12 (1927), pp. 1461–7.

[11]   Thomas Graham also developed practical laboratory exercises but it is generally agreed that this development was independent of Liebig.

[12]   James Muspratt, the Liverpool alkali manufacturer, developed a friendship with Liebig that embraced their families, and James's sons, who trained with Liebig, continued friendships with Liebig's children after the death of the two fathers.

[13]   W.H. Brock (ed.), *Justus von Liebig und August Wilhelm Hofmann in ihren Briefen* (Weinheim, 1984), p. 179.

a view to producing artificial manures that would increase the productivity of the soil. This was during a period from the late 1790s when Thomas Malthus's views on population growth and the ability to feed everyone became a major concern of those engaged in politics and theories of economics: 'The power of population is indefinitely greater than the power in the earth to produce subsistence for man'.[14] Liebig's work over many years was a direct challenge to such a view. But his concerns went wider, for he saw science and particularly chemistry leading the way in bringing benefits to all in society and overcoming the harsh and oppressive living conditions that so many found themselves in. This acted as a carrion call for many of Liebig's students, and for Angus Smith and for Lyon Playfair it was to become the *raison d'être* underpinning their future commitments to improving public health.

Influence of Meetings of the British Association for the Advancement of Science (BAAS)

The BAAS was formed in York in 1831. Its objective was 'to give a stronger impulse and a more systematic direction to scientific enquiry'.[15] The BAAS met annually in a different town with the aim of bringing together prominent natural philosophers (from home and abroad) with those in the local area to review recent advances in science and to address opportunities and concerns in the application of science. The work of the BAAS was conducted through different subject sections, with a president and secretary appointed to direct and organize each section, especially its programme for the annual meeting. From the BAAS's inception the chemistry section B was one of the most prominent sections.

It is through the annual meetings of the BAAS that we can see the network of chemists – Graham, Liebig, Playfair and later Angus Smith – working so effectively to advance chemical knowledge and promote its role in improving the conditions of society. It was as much through the annual meetings of the BAAS as it was through British students studying at Giessen that Liebig had such a powerful influence on British science.

The first BAAS meeting Liebig attended was in Liverpool in 1837 following an invitation from Thomas Thomson, Professor of Chemistry at the University of Glasgow, whose son had studied at Giessen. Accepting the invitation allowed Liebig to become reacquainted with many of his former Giessen students. Thomson sent his son Thomas to meet Liebig in Hull. Liebig was a seasoned traveller across Europe and overseas, and not daunted by the demands of travelling via York, Leeds and Manchester en route to Liverpool. In Manchester, Liebig's host for a

---

[14]   Thomas Malthus, *An Essay on the Principle of Population* (London, 1798), p. 13.

[15]   'Objects and Rules of the Association', *British Association Report*, 1842, p. v. See also Jack Morrell and Arnold Thackray, *Gentlemen of Science: The Early Years of the British Association for the Advancement of Science* (Oxford, 1981).

week was Charles Henry, son of William Henry, the first British student to study at Giessen, before going on to Liverpool to sail to Dublin, where he met another two former students, Robert Kane and William Gregory. Kane (later Sir Robert) had a distinguished career in Ireland, became the first President of Queen's College, Cork in 1845 and published two important and influential books, his three-volume *Elements of Chemistry* (1841–44) and *The Industrial Resources of Ireland* (1845). Gregory had an outstanding career in Scottish academia, becoming Professor of Chemistry at Edinburgh in 1844. He translated and edited some of Liebig's works, including *Animal Chemistry or Organic Chemistry in its Application to Physiology and Pathology* (1842) and *Research on the Chemistry of Food* (1847). Liebig then travelled to Glasgow via Belfast, where he met Thomas Andrews, who was to become Vice-President and Professor of Chemistry at Queen's University of Belfast in 1845.[16] In Glasgow, Liebig was hosted by the Thomsons (father and son) before once again meeting Thomas Graham (they had first met in Germany the year before), who had organized a tour of some industrial premises. The visit was brought to a fitting end with a farewell dinner for Graham following his appointment at University College London, at which Liebig was guest of honour.

As with other BAAS meetings, the Liverpool meeting proved another lively gathering, especially with Liebig attending. Michael Faraday, Professor of Chemistry at the Royal Institution of Great Britain in London, and one of the outstanding experimental scientists of the time, was President of the chemistry section. The section had invited Liebig to participate in the Liverpool meeting because of his interest in organic chemistry (then a developing field of enquiry). Liebig had prepared a paper on 'On the decomposition products of uric acid', but because his English was poor, Faraday read it. Because of the impact of his paper and his convivial disposition, Liebig became 'the lion of the evening' and was much sought after by delegates for his expert views. Sometime during the 1837 meeting, the BAAS council asked Liebig to prepare a report on the current state of organic chemistry. Knowledge of organic chemistry in the 1830s was rudimentary but Liebig had begun to use his system of analysis to elucidate the composition of organic substances and to apply the knowledge of these substances to advance medicine, pharmacy and agriculture. When Liebig returned to Giessen he had been away for two months and, as William Brock has summarized, the visit

> was of great importance in Liebig's intellectual development. He consolidated friendships with former British pupils who were themselves becoming well known, he forged important new friendships with, in particular, Graham, Crum, Muspratt, and Faraday; he increased Thomson's and Graham's desire to send their best pupils to Giessen for further training in organic analysis; and he had

---

[16] For fuller details of the tour, see W.H. Brock, *Justus von Liebig, The Chemical Gatekeeper* (Cambridge, 1997), pp. 96–7.

seen the advanced state of British industrialization and the high standard of living of its middle classes.[17]

For Liebig, the 1837 BAAS meeting cemented his influence on British science and he ensured that his published works were available in English. For the British, Liebig was to be honoured in 1840 when he was elected a Foreign Fellow of the Royal Society and awarded its Copley medal, and in 1841 when he was made a Foreign Member of the Chemical Society by means of Graham's influence. It was not Liebig's last contribution to a BAAS meeting.

The 1840 BAAS meeting in Glasgow was important for Lyon Playfair, who, although still a student at Giessen, was elected Secretary of the chemistry section and acted as surrogate for Liebig, who at this time was completing his work on agricultural chemistry. In the lead-up to the meeting Liebig had asked Playfair to translate into English his treatise on agricultural chemistry so that the ideas could be shared readily with his English friends and supporters. At the meeting Playfair not only delivered Liebig's talk on poisons and spoke about his own work on fatty acids, but developed professional friendships that were to influence his future standing in British science and, more importantly at the time, help him gain employment in the calico-printing industry in Clitheroe, Lancashire, as I shall discuss later in this chapter.

With the 1842 BAAS meeting taking place in Manchester, Playfair was again in a good position to take advantage of an opportunity to impress not only the local audience but also prominent members of the British scientific community. He continued as Secretary of the chemistry section, gave two talks on his own work and presented an abstract of Liebig's report, *Organic Chemistry applied to Physiology and Pathology*, requested by the BAAS council.[18] Playfair's talks were well received and led to discussion of topics that demonstrated his wide scientific knowledge and understanding. As Robert Kargon has pointed out, 'he became somewhat of a scientific celebrity', and this profile led to Playfair being offered a post in academia in Manchester.[19] From this time onwards he began to take a prominent and influential role in British science, frequently called in to advise Parliament and government on scientific matters and appointments, and to all intents and purposes he played the unofficial role of 'Scientific Adviser to the Government'. Playfair was to play a leading part in the 1851 Great Exhibition in London (alongside Prince Albert) and always took an active interest in scientific education when reforms were under review. It is very likely that Playfair had some part in the appointment of his friend and former colleague Angus Smith as the first Inspector of the Alkali Inspectorate in 1864.

---

[17]   Ibid., p. 97.

[18]   Kargon, *Science*, p. 90.

[19]   Ibid.

The Manchester Circle

The BAAS meetings in Glasgow and Manchester in 1840 and 1842 respectively gave Lyon Playfair not only a high profile within the scientific establishment, but also a network of valuable contacts that led to his appointment in two important positions – manager of a calico-printing firm in Clitheroe from the 1840 meeting and Professor of Chemistry at the Royal Manchester Institution from the 1842 meeting. Playfair's relocation to Lancashire provided the opportunities to work again with his close friends Angus Smith and James Young, while allowing him to actively promote and foster an interest in science within the local communities.

Following the Glasgow meeting, early in 1841, James Thomson, of James Thomson and Brothers, calico printers of Clitheroe, had consulted Graham and Liebig about a suitable candidate for the manager position at the dye-works.[20] Thomson had studied in Glasgow in the mid-1790s but then moved to London the following year to work for his cousin's firm, Joseph Peel and Company, and while there he developed a network of contacts that included Humphry Davy, William Wollaston, Gregory Watt (son of James Watt), Samuel Taylor Coleridge and Robert Southey. Thomson also undertook his own research, publishing two papers on dye-related matters in scientific journals; he was elected a Fellow of the Royal Society in 1828 and was a member of the Chemical Society at its inception in 1841.[21] When Thomson started searching for a new manager he wanted to appoint a chemist of some standing, not just to manage the dye-works but also to undertake research and development that might lead to valuable patents, just as Thomson had done in 1813 and 1816, and in these circumstances it is not too surprising that Thomson should seek advice from such eminent figures in chemistry as Graham and Liebig.[22] Their recommendation was Lyon Playfair, who was duly approach about the post and he readily accepted.

Besides his work for the Thomson company, Playfair undertook another venture in his spare time that would demonstrate his commitment to promoting and nurturing an interest in science among the working population. He used his home in Whalley (a village in the Calder valley) as a meeting place where a small group of workers in the nearby calico printers and chemical firms could share current scientific ideas (predominantly chemical) and the excitement of making new scientific discoveries. Each meeting focused on a topic, with one member of the group (in rotation) expected to give a presentation followed by a group discussion. Among the attendees were John Mercer and very probably James Young (his friend from Glasgow days).

---

[20]   Reid, *Lyon Playfair*, p. 44.

[21]   Kargon, *Science*, pp. 88–9.

[22]   Ibid., p. 89. James Thomson's patents were 1813/3654 (*Producing patterns on cloth previously dyed Turkey-Red, and made of cotton or linen, or both*) and 1815/3881 (*Process for printing cloth made of cotton or linen, or both*).

Figure 2.3   Lyon Playfair, 1st Baron Playfair (1818–1898), politician and chemist, lithograph by Thomas Maguire (1851)
Wellcome Library, London

Mercer was a partner at the calico printers Fox and Brothers.[23] In 1844 he developed the treatment of cotton that became known as 'mercerization', in which

---

23   P.J. Hartog (rev. W.H. Brock), 'Entry for John Mercer', *ODNB*.

sodium hydroxide strengthened the fibres and made them easier to dye.[24] The Whalley meetings were important for the development of Mercer's ideas:

> It is at these meetings that Mercer's theoretical interest flowered, and his interesting work on catalysis, which Playfair later extended, originated there. For men like Mercer the Whalley meetings provided their first and exciting encounter with the wider world of chemical research. As a recent Liebig student and Giessen Ph.D., Playfair brought to them the exhilaration of what they took to be research at the chemical frontier.[25]

James Young was born in Glasgow in 1811 and became interested in chemistry from 1830 through attending Graham's lectures at Anderson's University where he befriended Playfair and David Livingstone.[26] Young quickly developed a strong interest and ability in practical chemistry, and so impressed Graham that he became his laboratory assistant. When Graham left for University College London in 1837, Young joined him to continue as his laboratory assistant. In 1839 Young was appointed manager of the Muspratt works at Newton-le-Willows – after Muspratt had consulted Graham about a suitable person. This was during a period when Muspratt was expanding into the North American alkali trade while also preoccupied with trying to develop two new processes. One was an early form of the ammonia-soda process, while Gossage was struggling to get his sulphur-recovery process to operate efficiently on the large quantity of sulphur waste produced at Newton. In 1844 Young left Newton to become manager of the Ardwick Bridge chemical works in Manchester, part of Tennant, Clow & Company, and while there he took an active part in Manchester politics and was a prominent member of the MLPS as a close associate of Robert Angus Smith.

Events following the 1842 BAAS meeting in Manchester involved Playfair directly and Angus Smith indirectly in the affairs of the Royal Manchester Institution (RMI). The RMI had from its inception provided courses in science and literature through lecture series, but in 1843 the RMI Council decided to appoint two honorary professorships – in chemistry and physiology. James Heywood was a leading figure in the RMI, a Fellow of the Royal Society and founder member of the Geological Society, but more importantly for Playfair he was Local Secretary for the BAAS meeting. In this role Heywood had working contact with Playfair and was no doubt impressed with Playfair's contribution as Secretary of the chemistry section and his general contribution to the proceedings. It is very likely that Heywood's lauding of Playfair within the RMI Council resulted in Playfair being offered the position of Professor of Chemistry in 1843. What other forces

---

24   This treatment was improved in 1890 by H.A. Lowe to give the fibres greater lustre.

25   Kargon, *Science*, pp. 89–90.

26   John Butt, 'Entry for James Young', *ODNB*. See also Anthony Slaven, 'Entry for James Young', *Dictionary of Scottish Business Biography 1860–1960,* and John Butt, 'James Young, industrialist and philanthropist', PhD diss., University of Glasgow, 1964.

were at play is difficult to gauge: Playfair had previously been approached about becoming Professor of Chemistry at the newly established University of Toronto, much to the concern of the British science and political establishments at the prospect of losing such an outstanding scientist.[27]

Playfair's acceptance of the post at the RMI led in turn to the appointment of Angus Smith as his assistant. Their collaboration, with a report on the Large Towns of Lancashire as part of the Royal Commission on the Health of Towns, would exercise a huge influence on the future direction of Angus Smith's scientific work. Manchester would also provide Angus Smith with his future working base and many opportunities to contribute to the administration and proceedings of the MLPS. Angus Smith's association with Manchester is discussed in Chapter 3.

---

[27]   For a fuller treatment of this episode, see Kargon, *Science*, p. 91, and Reid, *Lyon Playfair*, pp. 58–61.

Chapter 3
Robert Angus Smith's Manchester

Manchester in the 1840s

Manchester became Angus Smith's working and living base from 1843 when Lyon Playfair appointed him as personal assistant at the RMI. Angus Smith remained there for the rest of his life, in preference to London or any other major city, even when he was appointed Inspector of the Alkali Inspectorate. Manchester was transformed by mass industrialization from the late eighteenth century, and by the 1840s was a town known by the sobriquet 'Cottonopolis'; this epitomized the good and bad effects of industrialization. Probably without knowing it at the time, Angus Smith had moved in 1843 to a town that would provide not only work for an analytical and consulting chemist, but also a 'laboratory' in which to investigate both the causes and amelioration of the insanitary conditions and from which he was to make his reputation as an environmental chemist and civil scientist. But how had Manchester grown into this major industrial and commercial town?

From the early stages of industrialization, Manchester had been the commercial centre of the region. It had become the centre of the cotton industry, moving from the hand-looms of a home-based trade to the steam-driven mechanization of a large factory-based industry serving the demands of the world. The factory production units were possible because of important inventions associated with three inventors – James Hargreaves (spinning jenny in 1764), Richard Arkwright (spinning frame in 1768) and Samuel Crompton (spinning mule in 1799) – and the ability to harness steam power to drive these machines following the innovations of James Watt. Together, these developments produced larger quantities of cotton thread of unprecedented quality and consistency. The outcome: images of Manchester from the early decades of the nineteenth century that depict the town as clusters of factories with their tall chimneys belching out black smoke from burning coal for their steam power.

While the cotton industry was no doubt an important factor in Manchester's growth as an industrial town, Alan Kidd has rightly drawn attention to another factor, one that pre-dates industrialization, and this is the role of the vast warehouses. He has even gone so far as to give the town another sobriquet, 'warehouse town', because of the part played by warehouses in the commercial development of Manchester, providing not only storage space but also areas for the display and sale of goods and for trading offices.[1] In terms of the relative importance to the development of Manchester,

---

[1]   Alan Kidd, *Manchester: A History* (Lancaster, 2006), pp. 16 and 20.

Total capital investment in factories was considerably less than in warehouses. Warehouses absorbed over 48% of property asset investment by 1815 as opposed to a mere 6% in factories. Even public houses and inns attracted a larger proportion at almost 9%. This does not allow for machinery or the power to drive it. But even assuming a doubling of value of the fixed assets of factory plant and buildings, the dominance of warehouse investment remains clear. Investment in cotton mills increased, especially as weaving was mechanised in the 1820s, but Manchester's business structure still leaned heavily towards its commercial sector. Whilst the proportion of all capital tied up in cotton factories had increased to some 12% by 1825 that invested in warehouses remained much higher at nearly 43%. Industrial Manchester was not a factory town which became a commercial centre; from the beginnings of industrialisation it has been a warehouse town with factories.[2]

However, focusing on factories and warehouses to depict the growth of Manchester does not give the full picture because Manchester was also an important centre of engineering. The engineering sector of millwrights, iron foundries, forging works and machine manufacture grew alongside the demand for machinery for the cotton factories. Manchester benefited from engineers such as William Fairbairn, James Nasmyth and Joseph Whitworth, among others. William Fairbairn (1789–1874) launched his mill-engineering business with James Lillie in 1817, and over time they became bridge builders, ship builders, locomotive manufacturers and inventors of the Lancashire boiler. James Nasmyth (1808–90) is best known as the inventor of the steam hammer but he also set up a machine-tool business. Joseph Whitworth (1803–87) is remembered for his invention of the screw thread but was also well known as a high-precision machine-tool engineer. During the nineteenth century they together forged Manchester into a pre-eminent engineering centre that added further to the perception of Manchester as an industrial and commercial regional metropolis.

For most major towns, the 1835 Municipal Corporation Act brought freedom from the feudal approach to governance of members re-electing themselves and pursuing their self-interest. The Act imposed regular elections that would engage with a wider group of the local citizenry and draw in the middle class to assist with governing the town. But because Manchester was not a corporation, it was excluded from the terms of the 1835 Act. A response soon followed. Manchester had always been a focal point of radical political action, and a campaign to have Manchester incorporated was initiated by Richard Cobden, the Manchester calico printer turned radical politician, with his tract, *Incorporate Your Borough – By a Radical Reformer*. The tract became a rallying point for the middle class with its no-nonsense approach to fighting the existing feudal governance in which the

---

2   Ibid., pp. 16–17.

Landlord interest used to make excursions from their strongholds to plunder, oppress, and ravage, with fire and sword, the peaceable and industrious inhabitants of the town.[3]

The campaign was successful and led to the incorporation of Manchester in 1838. But Cobden, having honed his political activism, decided to campaign for repeal of the Corn Laws, introduced in 1815 at the time of the Napoleonic Wars to restrict the importation of wheat. This drove up the cost of bread, a staple food for so many, in particular the poor in society. In 1839 Cobden became leader of the Anti-Corn Law League and seven years later managed to persuade the Prime Minister, Sir Robert Peel, to repeal the Corn Laws: 'The men of Manchester had brought down the nobility and gentry of England in a bloodless, but decisive Crécy'.[4] To commemorate this successful campaign, the Free Trade Hall in Manchester was constructed.

Good transport links were vital to a thriving Manchester. Until 1830 goods were transported by canal and passengers by coach. Most of the cotton for Manchester was imported through Liverpool, another thriving town, possessing a port connected to all parts of the world. Transport of cotton from Liverpool relied on rather slow conveyance on the River Mersey and the Irwell Navigation, but in 1823 a railway connecting the two towns was promoted. The chosen 64-mile route involved major engineering work, including some 66 bridges and viaducts, several long tunnels and a robust and safe passage over the bog of Chat Moss. Nevertheless, the railway line opened on 15 September 1830 with George Stephenson's *Rocket* locomotive at the head of the train. It became the first double-track inter-city passenger railway in the world.[5] The Liverpool and Manchester Railway was to start a railway-building mania across Britain over the next two decades that increased capacity for conveying both passengers and freight.

Manchester has always been associated with invention and enterprise, and as historian Robert Kargon has pointed out, 'the wealth and prestige of the 18th-century town [Manchester] lay with its clergy, professional men, and merchant-manufacturers'.[6] This led to the establishment of a number of influential cultural organizations, perhaps the most prestigious being the Manchester Literary and Philosophical Society (MLPS) established in 1781 'to facilitate the dissemination of literary and scientific culture among those gentlemen best prepared to make use of it'.[7] It was to this Society that Angus Smith devoted so much of his time and energy, as will be discussed later.

---

[3] Richard Cobden, 'Incorporate Your Borough', in W.E.A. Axon, *Cobden as a Citizen* (Manchester, 1907), p. 31.

[4] Tristram Hunt, *Building Jerusalem* (London, 2004), pp. 149–50.

[5] Unfortunately, the inaugural journey was also famous for the fatal accident involving William Huskisson. See Simon Garfield, *The Last Journey of William Huskisson: the day the railway came of age* (London, 2002).

[6] Kargon, *Science*, p. 3.

[7] Ibid., p. 4.

Other organizations were founded to cater for different sectors of society. The RMI was founded in 1823 to provide an art gallery in which to display fine art and offer meeting facilities for presentation of scientific lectures.[8] As discussed in Chapter 2, James Heywood was an influential member of the council but so too was George Wood, a leading merchant in Manchester, who was also associated with a number of institutions in the town as well as in London:

> His scientific interests were wide: he was a fellow of the Linnaean and Geological Societies, a member and vice-president of the Manchester Lit. and Phil. (a member of its council since 1810), and a moving figure in the founding of the Royal Institution and the Mechanics' Institution.[9]

It was the MRI that appointed Lyon Playfair as its first Professor of Chemistry in 1843, not long before Angus Smith joined him as assistant.

In 1824 the Mechanics' Institution was founded with the aim of providing lectures on the relationship between science and manufacturing, and with the intention of attracting the working class and artisans working in industry who wanted to further their understanding. By the 1840s these organizations were joined by a number of specialist subject societies that included the Manchester Natural History Society (formed in 1821) and the Manchester Geological Society (1838). The Manchester Statistical Society was founded in 1833 as one of the earliest provincial statistical societies, and became part of a wider national movement to use statistics to advance the social sciences and social policies, in much the same way as numerical data from physical experiments had influenced the progress of science and technology.[10] Many of the officers of these organizations and societies not only served their community in Manchester but also took an active part in London-based institutions, thus providing a useful bridgehead between Manchester and London, and bringing mutual benefits.

Another outcome of the growth of Manchester as a regional town over the first half of the nineteenth century was the dramatic rise in population: in 1801 the population was 76,788; by 1821 it had risen to 129,035 and by 1841 it was 242,983. While the population of London had doubled over the same period, Manchester's had tripled. As with so many large towns, especially in the north of England, a feature of this growth was that people in rural communities, invariably facing hardship, migrated to the new urban communities, attracted by work in the new factories or commercial enterprises. As Tristram Hunt has pointed out:

---

[8]   The building in Mosley Street is now occupied by Manchester Art Gallery.

[9]   Ibid., p. 19.

[10]   In 1834 the MSS carried out the first house-to-house survey. See Margaret Schabas, review of 'The Rise of Statistical Thinking 1820–1900 (Princeton, 1986)', in *Victorian Studies*, 31 (1987), p. 123.

Except maybe Oldham and Ashton, most industrial cities had a heterogeneous economic base which encompassed the commerce, financial, retail service, as well as industrial sectors. Manchester was as much dependent upon its mercantile base, construction industry and retail sector as its cotton mills. The 1841 census revealed that some 41,000 were employed in textiles in Manchester while domestic service gave jobs to 14,000 and the building and retail trades provided work for another 7,000 each. The city's wealthier citizens were more likely to be bankers, brewers or merchants than the mill-owner of popular myth.[11]

But there was an added component to this population growth for many large towns close to the west coast of Britain, principally Liverpool, Manchester and Glasgow: the many thousands who fled Ireland in the period leading up to 1846, suffering starvation due to the potato disease.. It was estimated that 40,000 Irish migrants arrived in Manchester, and by 1851 the Irish-born represented 13 per cent of the population of Manchester and Salford.[12] This influx over a relatively short period aggravated an already difficult housing situation since the migrants were forced into poor-quality housing such as back-to-back and cellar dwellings. These were very susceptible to the spread of diseases such as cholera and typhus, as we shall see later. This raises the question of the living conditions in Manchester in the 1840s and how they had come about.

Manchester's Air and Sanitation in the 1840s

Manchester epitomized a regional metropolis undergoing remarkable change through its industrial and commercial development. The adoption of steam power fuelled by continuous burning of coal enabled the factories to meet the increasing demand for cotton, and the Manchester landscape was characterized by a bewildering number of tall chimneys – the number providing a ready indicator of industrial output. But the chimneys were not benign instruments of industrialization; they almost continuously belched out black smoke from the inefficient burning of coal that often contained a high percentage of sulphur. This smoke was not the sole outcome of industrialization, for in 1661 John Evelyn wrote of his experiences in London:

> And what is all this, but that Hellish and dismall Cloud of SEA-COALE? which is not onely perpetual imminent over her head; For as the Poet,
>
> *Conditur in tenebris altum caligine coelum;*

---

[11]   Hunt, *Building Jerusalem*, pp. 15–16.
[12]   Ibid., p. 16.

but so universally mixed with the otherwise wholesome and excellent *Aer*, that her *Inhabitants* breathe nothing but an impure and thick Mist, accompanied by a fuliginous and filthy vapour, which renders them obnoxious to a thousand inconveniences, corrupting the *Lungs*, and disordering the entire habit of their Bodies; so that *Catharrs, Phthisicks, Coughs and Consumptions*, rage more in this one City, than the whole Earth besides.[13] [Italics in the original]

Evelyn was writing about London, referring particularly to the burning of coal by bakeries and in the domestic hearth; for towns affected by intense industrialization the smoke issue became ever more acute as the amount of coal burnt had grown rapidly from the mid- to late eighteenth century. The smoke was caused by inefficient burning of coal, so that instead of just carbon dioxide and water vapour as the main products, there was some carbon monoxide and soot (absorbing oily hydrocarbons) – the latter was responsible for blackening the stonework of buildings and also for creating the very dark clouds descending regularly over, say, Manchester even on a bright summer's day. Such an apocalyptic sight was probably the defining moment for Angus Smith as he walked into the town centre from his house in November 1844. It led to his subsequent investigations into the contents and causes of polluted air. He recorded his shock and lasting impressions in a letter to *The Guardian*:

Coming in from the country last week on a beautiful morning, when the air was unusually clear and fresh, I was surprised to find that Manchester was enjoying the atmosphere of a dark December day. This is not seldom the case; we see the town filled with a dense vapour, and we persist in calling it fog; we see that the atmosphere of our streets in winter is frequently of the deepest black, such as the eyes cannot penetrate above a few hundred yards, and we persist in supposing that it is necessarily connected with the season, and with it only. But no such atmospheres are to be found far from large towns. There are times when the smoke ascends easily, and a slight breeze removes in an instant the accumulation of carbon over the town, when in fact no amount of smoke seems capable of greatly injuring the air; but such is the rare exception, and the rule is the thorough saturation of every street and house with the produce of our coal-pits and furnaces.[14]

It is significant to note as a further portent of the damaging conditions in Manchester that in 1848 the lepidopterist R.S. Edelston made the first recorded observation of the dark *carbonaria* form of the peppered moth.[15]

---

[13]   John Evelyn, *Fumifugium or The Inconvenience of the Air and Smoke or London Dissipated* (London, 1661).

[14]   *The Guardian*, 2 November 1844.

[15]   Stuart Hylton, *A History of Manchester* (Chichester, 2003), p. 149. The peppered moth with its lightly patterned wings adapted to a black wing morph that could survive

As Stephen Mosley has pointed out, in 1870 Manchester had 1,333 industrial works, with a further 296 in Salford, reflecting the wide range of industries from the period of unrelenting industrialization.[16] With the growing adoption of steam power to drive machinery, the number of chimney stacks increased, and by the 1840s there were about 500 chimneys in Manchester.[17] Such statistics were invariably used as a measure of industrial output and economic progress.

Industry's contribution to the black smoke was joined by the growing number of domestic hearths burning coal for cooking and heating. In 1866 Sir Robert Peel drew attention to the consumption of coal in Manchester, estimated at 2,000,000 tons per annum.[18] Such consumption in a relatively small area added to the insanitary and unhealthy living conditions. But the Lancashire coal burnt in Manchester contained up to 3 per cent of sulphur and added a further component to black smoke in its damaging effects on the local inhabitants, as will be discussed in more detail in Chapter 4.

It was not just choking air but also inadequate fresh water and poor disposal of sewage that created the insanitary conditions: sewage regularly polluted the water supply, increasing the spread of diseases such as typhus, cholera and diarrhoea. Cholera epidemics occurred in Manchester (as in many of the larger towns in Britain) in 1832, 1849, 1854 and 1866, and the poor housing and general living conditions contributed to their proliferation. Moreover, the housing stock was completely unable to adequately house the fast-increasing population. While many middle-class families lived in traditional terrace housing, those less well-off were forced to live in overcrowded back-to-back housing and in cellar dwellings. The back-to-back houses were

> built in double rows, often forming courts, each house having its only door and window at the front ... they were poorly ventilated (no back yard), and the water pump and privy (drained to a cesspit or ashpit) were shared, commonly between 20 houses.[19]

The cellar dwellings provided even unhealthier living conditions:

> These were often the residue of former grandeur, being higher grade housing which had declined and been converted into several tenancies occupying whole floors and single rooms, of which the meanest part was the cellar ... [they] were

---

predators when resting on soot-covered surfaces.

[16] Stephen Mosley, *The Chimney of the World: A History of Smoke Pollution in Victorian and Edwardian Manchester* (Cambridge, 2001), p. 17–18.

[17] Ibid.

[18] Ibid.

[19] Kidd, *Manchester*, p. 39.

often damp, dark and unhealthy. In the 1830s and 1840s up to 20,000 people existed in these almost subterranean dwellings.[20]

Local inhabitants probably got used to the unhealthy, stinking and murky conditions they found themselves in every day, but visitors to the town were shocked by what they found. Charles Dickens was a regular visitor to Manchester and it was probably his experiences there that led him to portray Coketown in *Hard Times* as

a town of red brick, or of brick that would have been red if the smoke and ashes had allowed it: but as matters stood, it was a town of unnatural red and black like the painted face of a savage.

It was a town of machinery and tall chimneys, out of which interminable serpents of smoke trailed themselves for ever and ever, and never got uncoiled.

It had a black canal in it, and a river that ran purple with ill-smelling dye, and vast piles of buildings full of windows where there was a rattling and a trembling all day long, and where the piston of the steam-engine worked monotonously up and down, like the head of an elephant in a state of melancholy madness.[21]

Foreign visitors such as the French traveller Alexis de Tocqueville were also shocked by the degree of squalor they found:

the greatest stream of human industry flows out to fertilise the whole world. From this filthy sewer pure gold flows. Here humanity attains its most complete development and its most brutish; here civilisation works its miracles, and civilised man is turned back into savage.[22]

When Angus Smith arrived in Manchester in 1843 he settled in the Chorlton-upon-Medlock district, where some of the harshest living conditions were experienced. The rise in population in quite a small district aggravated the already difficult situation, as a reported lecture (probably given in 1854/55) summarizes, drawing on information from the 1851 census:

[The district comprises] 700 statute acres, contains 6,951 dwelling houses, and a population of 35,546 individuals, of whom 16,272 are males and 19,274 are females. The population in 1831 was 20,569; in 1841, 28,322, so that in twenty years there has been an increase amounting to nearly 15,000 inhabitants.

---

20   Ibid.
21   Charles Dickens, *Hard Times* (London, 1995), p. 18.
22   Alexis de Tocqueville, *Journeys to England and Ireland*, trans. G. Lawrence and K.P. Mayer (London, 1958), pp. 107–8.

From the returns I have received from the police it appears that there are now 7,708 houses, and 39,962 people residing within the district.[23]

It was not just the dramatic increase in population but also the quality of the housing, particular the back-to-back and cellar dwellings, that were causes of alarm:

The third class [of dwelling] are those which are built back to back, in close confined situations, where frequently there is no thoroughfare, the end of the street being built up, and forming a sort of cul-de-sac.

These dwellings have no back yards or other such advantage, and the inhabitants had formerly but one place of convenience for about twenty houses. These tenements usually consist of two rooms, viz. a house-place and a bedroom; but in many, which are overcrowded, the former is also used as a sleeping apartment. Some of these houses, however, are three storeys high, with cellars beneath, in which case the upper rooms are sub-let to other families, and the cellars let off as separate dwellings. The latter are low, damp, without ventilation, almost destitute of light, and inhabited by the most improvident class of people.[24]

The houses were poorly constructed:

and when it is considered that the houses are built in such a loose slender style, with walls only of half-brick thickness, and boards half-an-inch apart, it is not surprising that deaths from these affectations should constitute so large an item in the registers of mortality.[25]

Another contributory factor to the environmental conditions in the area known as Chorlton-upon-Medlock was the River Medlock. The section of river running through the district was 2,300 yards long, and with its average width of 10 yards, created a polluting ditch of 23,000 square yards:

by [its] in-pouring of refuse has become one of the greatest nuisances in the township – in fact it is nothing better than an open ditch … If one takes the trouble to look at the Medlock, about Hulme-street, a number of dead dogs and cats may constantly be seen in the several stages of decomposition; … bubbles of gas, chiefly light carburetted hydrogen, rise to the surface, and although offensive smells are met with at all times, they are by far the more annoying

---

[23]   Lecture report, p. 7. Records of the Manchester and Salford Sanitary Association, GMCRO (Ref: M126/5/1/17).

[24]   Ibid., pp. 8 and 9. See also Robert Angus Smith, 'Ventilation and the Reasons for It', *Popular Science Monthly*, 1 (1872), pp 356–62.

[25]   Ibid., p. 13.

... Sulphuretted hydrogen is the gas which chiefly causes the odour, although doubtless phosphoretted hydrogen gas assists in some measure.[26]

It was such harsh living conditions that informed the 1845 report of the Health of Towns Commission (to which Angus Smith contributed), which led to the Public Health Act of 1848 with its provision for medical officers of health. They were to take the lead to ameliorate these unhealthy living conditions and prevent diseases such as cholera and typhus spreading so readily.

Lyon Playfair, the Royal Manchester Institution and the Health of Towns Commission

In 1843 Lyon Playfair was appointed Professor of Chemistry at the RMI. This reflected the growing importance of chemistry and the standing of the RMI within Manchester's cultural life at this time. Before his appointment Playfair had delivered a series of lectures that were well received, and following the announcement of his professorship his lecture programmes and laboratory proved very popular with students and also with John Dalton, famous for his atomic theory, who was a regular visitor. This support was important financially for Playfair, as Robert Kargon has pointed out:

> the professorship was itself unpaid. Playfair received, however, rooms which he fitted up as a laboratory and compensation for courses of lectures delivered through the year. The prestige of being the only 'professor' of chemistry in the city was a distinct advantage in drawing private pupils to his laboratory.[27]

As Playfair recalled, because there were so many students, 'I had to secure the services of assistants, one of whom was Dr. Angus Smith, so well known afterwards for his researches on air and disinfectants'.[28]

When the Health of Towns Commission was set up in 1843, Lyon Playfair was appointed one of the commissioners. Since he was based in Manchester he took responsibility for reporting on the Large Towns of Lancashire. Realizing the enormous amount of information this work would involve, he asked Angus Smith to become an assistant commissioner on the report.

For Angus Smith this appointment was to prove the final resolution of his personal struggles as to whether to pursue a career in the Church or in chemistry. Having worked with Playfair on the report for the Royal Commission on the Health of Towns from 1843, he was connected for the rest of his life with Manchester and

---

[26]   Ibid., pp. 10–13.
[27]   Kargon, *Science*, p. 91.
[28]   Reid, *Lyon Playfair*, pp. 56–7.

the issues associated with the insanitary conditions affecting so many of the larger towns in Britain at this time.

The report was published in 1845 as an Appendix to the Commission's Second Report. It is a thorough survey of the conditions in the large towns of Liverpool, Manchester, Preston, Bolton, Wigan, Ashton-under-Lyne, Bury and Rochdale; Playfair had intended including the towns of Oldham and Blackburn, but limitations of time prevented this. The report is divided into two parts:

> 1st. A consideration of the state of the towns as regards drainage, cleansing, supplies of water, building regulations, etc. According to returns made to me by local public bodies and acknowledged authorities in the several localities; and an examination of the various Local Acts and usages prevailing.

> 2nd. The consequences upon public health and morals, and the burdens on the community occasioned by the absence of sanitary regulations in the towns examined, as ascertained by inquiries instituted by myself with the aid of those conversant with the localities.[29]

Playfair and Angus Smith approached people in each town whom they knew would provide reliable and sufficiently detailed information against a standard set of questions. The headings in the report show the wide range of interests of the Commission: sewerage and drainage; scavenging; cesspools and necessaries [privies]; nuisances; streets and sites of houses; dwellings of the poor; lodging houses; public schools; public parks; public baths and laundries; supply of water; local acts and usages. The intention of the final heading was to review what the current local Acts were achieving, where the current application of the available Acts was aggravating the public health conditions rather than improving them and pointing out where further enhancement of the local Acts would improve public health. The information gathered about each town was then collated into a narrative with the supporting data tabulated.

This part of the Report was followed by a section on 'Physical Causes of Excessive Mortality', which tried to identify where new legislation was required to improve public health by drawing attention to the relation of births and deaths, general effects of causes of disease, administration of opiates to children, moral causes of disease. It included a supplement on 'The alleged causes of the excessive mortality in Liverpool, and on the fallacy that the excess of Mortality in a town is due to its migrant population'.[30] This pointed out that the census was undertaken during the period of the year when emigration was at its height and included those most likely to suffer disease and to die. There was also the difficulty of estimating

---

[29] Health of Towns Commission, *Reports of Commissioners, Appendix, Part II, Report on the Sanitary Condition of the Large Towns of Lancashire*, 1845, P.P. 1845 (602) (601), p. 1.

[30] Ibid., pp. 73–82.

the number of deaths among those returning immigrants who were often in a state of destitution. The final part of the Report included responses to 62 questions posed by Playfair (and Angus Smith) as part of their enquiries.[31]

Angus Smith's experience of working with Playfair on the Large Towns of Lancashire Report was another important milestone in determining the future direction of his professional interests. While he was becoming increasingly aware of the public health issues in Manchester, it was also clear that most of the issues were universal among most large towns, with the scale of the issues dependent on the size of the town. Even if legislation was forthcoming following publication of the Health of Towns Commission's Report, much remained to be done to understand scientifically the causes and how best to tackle the remedies if improvements in public health were to be long lasting.

Angus Smith as Consulting and Analytical Chemist

When Playfair decided to leave Manchester for London in early 1845 to take up the post of Chemist to the Geological Survey, Angus Smith was again left in one of his periodic quandaries about the future direction of his life, perhaps again even reflecting on the possibility of working in the Church. In the short period spent in Manchester Angus Smith had become acquainted with the harsh and unhealthy living conditions in the town, but more importantly for an aspiring consulting and analytical chemist, he had gauged the potential business opportunities outside academia, given the diversity of industrial interests in Manchester and the surrounding towns. Finally, he took the decision to remain in Manchester, where over the next 40 years or so he was to develop a national reputation as an outstanding analytical and consulting chemist.[32]

Angus Smith's work brought him in contact with a diverse group of industrial concerns that reflected the growing importance of chemistry, in particular analytical chemistry, in solving some of society's challenges, acting as a brake on the *laissez-faire* approach of industry, especially as far as pollution was concerned, and as a contribution to several major parliamentary inquiries that helped drive the remarkable reform legislation of the second half of the nineteenth century. These actions, together with those of many other chemists and scientists from other disciplines, would add to the professionalization of science during the same period.

Investigating the effective working of chemical processes in industry formed an important part of Angus Smith's work, but his main focus was on air and water

---

[31]   Ibid., pp. 84–118.

[32]   Angus Smith's laboratory was separate from his home. The addresses changed over time but from 1863 he lived at 5 York Place, Chorlton-on-Medlock, while his laboratory was at 29 Devonshire Street.

analyses, for which he would steadily build up a national reputation.[33] Railway companies, for example, were concerned about the quality of their water and in particular its degree of hardness because 'furring up' could lead to boiler malfunction or, worse still, an explosion, although such an extreme outcome was unlikely.[34] Surviving records show that Angus Smith carried out analyses on 12 samples of water for the LNWR Company in 1853, but it is very likely that he carried out many more, and for other railway companies.[35] Later, as analysis diversified into fuel analysis and preservation of timber, the railway companies found it more economic to set up their own laboratories and appoint their own chemists, although these were often trained by the consultants apppointed by the companies, given the increasingly specialist nature of the analyses.

Angus Smith carried out a number of water analyses for Edwin Chadwick, the social and public health reformer whose *Report on the Sanitary Condition of the Labouring Population of Great Britain* (1842), together with the Report of the Health of Towns Commission, was so influential in informing the Public Health Act of 1848 and the appointment of medical officers of health. It was probably while Angus Smith was working with Playfair on the Report for the Health of Towns Commission during 1844 and 1845 that he became acquainted with Chadwick and became a follower of his approach to sanitary issues. From the surviving correspondence it appears that Angus Smith and Chadwick were in regular contact from 1845 until the end of Angus Smith's life in 1884. While the correspondence relates mainly to sanitary issues and to analyses regarding water quality, there are references to travels (including a visit to Rome that demonstrated the merits of cement as a building material) that convey more of a personal friendship and not solely a professional relationship.[36]

Angus Smith's work on water and air quality became more widely known through his talks and articles for the Manchester Literary and Philosophical Society (MLPS) and his articles in publications of national and international renown. As a result, he steadily developed a profile of national importance. Later, through his work as Inspector (and later Chief Inspector) of the Alkali Inspectorate, he gained an international reputation. Besides the work on air and water quality that he felt passionately about, Angus Smith frequently acted as a consultant chemist, inspecting chemical works and reporting on how the operation of the works and its processes could be improved, giving expert evidence in court cases in which

---

[33] Unfortunately very few records of these analyses have survived, except some done for railway companies and some used for parliamentary inquiries (which were extensive).

[34] John Hudson, *Chemistry and the British Railway Industry 1830–1923*, PhD diss. (Open University, 2005), p. 19. See also Colin A. Russell and John Hudson, *Early Railway Chemistry and its Legacy* (Cambridge, 2012).

[35] LNWR Crewe, Register of Water Analyses 1853–1883, CRO (Ref: CCA/ NPR/3906).

[36] The Chadwick Collection is housed in the Special Collections Library, University College London.

chemical works were accused of creating a nuisance, and advising Florence Nightingale on how water quality management might be improved in India. There were contributions to a number of important government inquiries, including the Royal Commission on the Sanitation of the Metropolis (1848), the Royal Commission on Mines (1864) and the Cattle Plague Commission (1865–66). These were in addition to his work from 1864 as Inspector (later Chief Inspector) of the Alkali Inspectorate, his annual report to Parliament and his submissions to several parliamentary inquiries to review the working of the extant legislation and identify where tighter controls and extension of the regulations to other industrial sectors was felt necessary to reduce pollution. These issues are discussed in more detail in later chapters.

Manchester Literary and Philosophical Society

Angus Smith was elected a member of the Manchester Literary and Philosophical Society (MLPS) on 29 April 1845, perhaps with enthusiastic support from Lyon Playfair, who, although not resident in Manchester at the time, had been a member since 1842. For Angus Smith, the MLPS was to become his 'substitute for academia' and the main focus of his wide scientific and cultural interests. He applied unsuccessfully for the post of Professor of Chemistry at Owens College in 1850 (Edward Frankland was appointed) and in 1857 (lost out to Henry Roscoe), and in 1862 was unsuccessful in his application for the Professor of Chemistry post at the University of Aberdeen.[37] Except for membership of the specialist chemical societies, he refrained from taking an active part in London-based societies and focused almost exclusively on the MLPS. Not only did Angus Smith contribute many talks at the MLPS's regular meetings, and publish many articles in the Society's printed proceedings, but he gave unstinting service as an outstanding officer in several capacities over an extended period of time.

The MLPS was founded in 1781 and it is generally accepted that Dr Thomas Percival, a Manchester surgeon who had trained in Edinburgh and Leiden, was the main promoter. At its inception the Society's ethos was a club for gentlemen with the aim of diffusing knowledge and encouraging discoveries. Topics for discussion were wide and varied, but any relating to politics, medicine and religion were not permitted. The number of members was restricted to 50; the membership was allowed to build up gradually to this number over the years through a ballot system. The MLPS became Manchester's leading scientific institution over time. The first volume of the *Memoirs* was published in 1785 and papers for inclusion in the Society's publications were controlled by a Committee of Papers, a rather quaint name for an editorial board although it allegedly had draconian powers.

---

[37]   *Testimonials in favour of Robt. Angus Smith, Ph.D., F.R.S., F.C.S.,...candidate for the Chair of Chemistry in the University of Aberdeen* (London, 1862).

No one better represents the leading members of the Society in the 1780s than Thomas Henry, of the well-known Henry family of Manchester. He was an apothecary and chemical manufacturer in the town and a Fellow of the Royal Society who promoted the MPLS as an institution for the study of literature and science by the 'amateur', although he stressed the importance of science for local merchants and industrialists, in particular the benefits of a working knowledge of chemistry. Remarkably, Henry was an officer of the Society almost continuously from its inception until his death in 1816.

Another scientist who was to have an even greater influence on the Society was John Dalton. Dalton was elected a member in 1794, an officer in 1800, and he was to be a towering figure in the Society and in the Committee of Papers. With Dalton's involvement in the affairs of the Society, there was a move away from the 'amateur' interest and, as Robert Kargon has highlighted:

> In May of 1800 Dalton was elected secretary of the society, a task which entailed a fair amount of correspondence and society business. He held the post until 1809 when he was elected vice-president, a position to which he was re-elected each year until his succession to the presidency in 1817. From 1817 until his death in 1844 Dalton presided over, and to a large extent remade, the image of the society. The claim – often made – that owing to Dalton's preferences the society increasingly emphasized science at the expense of other studies is difficult to assess ... under Dalton about sixty-five to seventy-five percent of the papers published in the Lit and Phil Memoirs were scientific, but this reflects a trend apparent already in the 1790s.[38]

When Angus Smith joined the Society in 1845, just after Dalton's death, he would have been aware of Dalton's influence in setting the Society on a course towards a 'professional scientific society'. It is quite likely that this reference to 'scientific' attracted Angus Smith and the many other scientists and engineers who became members and took part in its activities, such as Lyon Playfair, James Young, William Fairbairn, James Joule, Edward Schunck, Frederick Crace-Calvert, Alexander McDougall and Peter Spence, among many others. Moreover, it is probably this exciting intellectual environment within the MLPS, created by these outstanding scientists and engineers, that precluded any ready inclination to engage with London-based societies. For Angus Smith, the MLPS's meetings and publications provided the principal outputs for both his scientific work and his wide cultural interests.

As did John Dalton, Angus Smith took on several important officer roles over very many years. Just five years after becoming a member, in 1852 he was first elected Secretary and occupied this post until 1856 (in 1856 jointly with his friend and fellow student from Giessen, Edward Schunck). Then in April 1859 he was first elected Vice-President and retained this position almost continuously (in

---

[38]   Kargon, *Science*, p. 12.

many years jointly with Edward Schunck) until 1878. He had the honour of being President of the Society in 1864 and 1865. It is interesting to note that Angus Smith held several of these positions while Inspector of the Alkali Inspectorate, and that while he was Vice-President, Thomas Graham and Wilhelm Hofmann, Director of the Royal College of Chemistry in London, were elected honorary members. In 1863, while Vice-President, Angus Smith was involved in an unfortunate dispute over the election of a new member that escalated into the question of whether the Society should have wider public appeal but in the end the Society determined to retain its science and literary focus.

Edward Schunck was a close associate of Angus Smith's within the MLPS. Although he had been born in Manchester, Schunck studied chemistry in Germany, including a period with Liebig at Giessen, where he was awarded a PhD and befriended Angus Smith. He returned to Britain to work in his father's calico-printing works but his main interest was research into various colouring agents, including madder and indigo, building a well-equipped private research laboratory with an extensive library.[39] Schunck was elected a member of the MLPS in 1842, was Secretary between 1855 and 1860, and was President in 1866–67, 1874–75, 1890–91, 1896–97. He was the first recipient of the Society's Dalton bronze medal in 1898. He was also an active member of the Chemical Society (1841) and the Society of Chemical Industry (1881). In 1850 he had been elected a Fellow of the Royal Society and in 1899 received the Davy gold medal. Nevertheless, Schunck, like Angus Smith, always retained Manchester as his home and working base, and gave unstinting commitment over many years to the affairs of the MLPS.[40]

Angus Smith's many talks and articles for the MLPS are discussed in more detail in subsequent chapters and are summarized in the Robert Angus Smith bibliography.

Manchester and Salford Sanitary Association

In October 1852 the Manchester and Salford Sanitary Association (MSSA) was formed. Among the promoters were several members of the MLPS who shared concerns about the insanitary conditions in the two towns, including Angus Smith and Frederick Crace-Calvert. The objective was

> To promote attention to Temperance, Personal and Domestic Cleanliness and to take the Laws of Health generally ... to induce general co-operation with the Boards of Health and other constituted authorities in giving effect to official Regulations for Sanitary Improvement.[41]

[39] 'Edward Schunck', *Journal of the Society of Chemical Industry*, 50 (1931), p. 65.

[40] Anthony S. Travis, 'Entry for Edward (Henry) Schunck', *ODNB*.

[41] Administrative History, Manchester and Salford Sanitary Association, GMCRO, see entry on website www.a2a., p. 2.

And the work of the MSSA was organized whereby

> Manchester and Salford were divided into seven districts – Deansgate, London
> Road, Ancoats, Oldham Road, Rochdale Road, Chorlton and Salford – and
> committees of visitors appointed to make personal investigation of the sanitary
> conditions in houses, courts and streets of their district.[42]

The aim in founding the Association was not to replace the statutory bodies
responsible for improving the sanitary conditions but to provide support through
surveys and talks within the different community areas comprising Manchester and
Salford. Surveys were concentrated in areas where the problems were most likely
to be acute in order to focus the amelioration work of the statutory authorities. The
other core activity for the Association was the programme of regular public talks
held in church halls and other community meeting places to emphasize the role
that personal responsibility could play in improving public health.

The Association brought together many prominent local people who were not
all members of the MLPS or chemists with an interest in sanitary matters, though
many were. They included, as vice-presidents, the engineers William Fairbairn
and Joseph Whitworth. The Association had a chemical and geological committee
whose members included Angus Smith, Frederick Crace-Calvert, Alexander
McDougall (a prominent member of the MLPS and a future collaborator with
Angus Smith on a disinfectant powder), Daniel Stone (a local consulting chemist),
Edward Binney (Secretary of the Manchester Geological Society), James Allen (a
chemist who knew Playfair and had trained with Liebig in Giessen) and Edward
Frankland (Professor of Chemistry at Owens College from 1850 to 1857).[43]
This remarkable group of people, representing many different disciplines, came
together in their shared belief that through group effort they could help improve
the appalling living conditions existing at the time.

Two reports will illustrate the types of investigation carried out on its behalf. In
October 1853 Angus Smith, Crace-Calvert and Stone investigated the state of the
water supplied by the Manchester Water Works to the inhabitants of Manchester and
Salford. Even though an epidemic of cholera threatened the towns, this study was
recognized from the outset as a progress report to compare the current quality of the
water with previous supplies that were found to be contaminated with 'animalcules'
because the water was drawn from agricultural land. The report concluded 'that the
water supplied to the districts in which our laboratories are situated has been quite
incapable of putrifying [sic] since we were requested to make our report'.[44] This
improvement was very welcome and reassuring since good water quality had a
crucial part to play in public health. With water demand rising steadily through the

---

[42]   Ibid.

[43]   Kargon, *Science*, p. 123.

[44]   *First Report. Appendix No: III* (AGM on 8 November 1853). Records of the
Manchester and Salford Sanitary Association. GMCRO (Ref: M126/5/1/15).

rest of the nineteenth century, Manchester Corporation completed the Longendale Reservoir (in the 1850s) and Thirlmere Reservoir in the Lake District (in 1894). By 1900 water consumption had risen to nearly 32 million gallons a day, a level of demand that led to construction of the Haweswater Reservoir in 1920.[45]

The second report required Angus Smith and McDougall to carry out experiments on 'a Method of Purifying the River Medlock and the Bridgewater Canal'. The river and canal were connected and formed one of the main water arteries for Manchester, but it was heavily polluted by industrial waste and also domestic sewage since Manchester had shown little interest in adopting the water-closet system:

> The water of the Medlock, as is well known, is of dark bluish tinge, somewhat resembling ink at a distance, and ceasing to be transparent when more than an inch deep. [46]

In their experiments Angus Smith and McDougall were interested to discover the amount of lime needed to fully treat and deposit all the 'carbonic acid', organic matter and other substances held in solution and suspension in the water. They constructed a settling tank and tried to ascertain how rapid the current of water should be to allow deposition of the precipitate when the lime was added. They had started with a flow of 125,000 gallons per hour but this had to be reduced to 41,000 gallons per hour to achieve full deposition of the precipitate. Although they found the lime/precipitate could be re-used on four occasions, with the flow of the Medlock at about 15,000,000 gallons per day approximately 12 tons of lime would be required and this was completely out of the question because of availability and cost. They came to the conclusion that a large depositing reservoir was probably not the correct approach; the water had to be purified before it entered the Medlock and further consideration should be given to adopting water-closets to control domestic sewage. Investigations such as these two examples allowed for some serious experimentation before committing to expensive major engineering schemes.

Another key part of the Association's activities was keeping up-to-date records of the conditions in each of the districts across the two towns as defined by the Association. Each district set up a 'Visiting Committee' to carry out regular surveys in the houses, courts and streets and that could report urgently on recent dangerous episodes. To ensure thoroughness and consistency across the different 'Visiting Committees', the Association adopted an assessment pro-forma for completion during the inspections.[47]

---

[45]   Kidd, *Manchester*, p. 126.

[46]   *Minutes of Sub-Committee on Sewer Rivers, Second Annual Report, 1856*. Records of the Manchester and Salford Sanitary Association. GMCRO (Ref: M126/5).

[47]   Records of the Manchester and Salford Sanitary Association, GMCRO (Ref: M126/5/1/23).

The programme of public talks in the different community locations across Manchester and Salford formed another vital element of the Association's work. Surviving leaflets indicate the range of these talks, aimed at promoting personal responsibility for improving public health. Angus Smith, as Chairman of the Chorlton District Committee, spoke on themes such as: 'Atmospheric Air, its Uses, the Sources of its Deterioration, and subsequent Prejudicial Effects' (12 January 1854); 'Air, pure and impure, and its relation to Health' (16 February 1854); '*Food*, and the Changes it undergoes in the Digestion Organs, and how Health is secured or Disease Engendered by Food and Drink' (15 February 1855); 'On Atmospheric Air' (21 February 1855); 'Habits of Order observable in Nature' (9 March 1857 and 4 March 1858); 'The Economy of Cooking' (12 February 1858).[48] Unfortunately, no summaries have survived, but in the case of his last lecture, it is interesting to speculate whether Angus Smith drew on his own cooking experiences or took a more theoretical approach.

Another person to take a leading role in the activities of the Association was Frederick Crace-Calvert, a friend of Angus Smith's and a collaborator with him on several occasions. Crace-Calvert received his chemical training in France working with Michel-Eugène Chevreul at the Gobelins dye-works and the Musée d'Histoire Naturelle between 1841 and 1846.[49] He was attracted to Manchester on his return to Britain and later in 1846 was appointed Professor of Chemistry at the RMI to replace Lyon Playfair. He started investigating the production of phenol for making dyes and as a disinfectant, and this led Crace-Calvert in 1859 to set up a manufacturing company for phenol and related products. As J.R. Crellin has pointed out:

> It was Crace-Calvert's pure phenol that fostered Joseph Lister's far-reaching work on antiseptic surgery – Lister used phenol to destroy micro-organisms associated with post-operative infections. Lister's work was first published in 1867, but even before then Crace-Calvert had been to the forefront in promoting sanitary and medical uses of phenol, as disinfectant for sewage and in the management of sloughing wounds, foetid ulcers, and diarrhoea.[50]

Crace-Calvert was a member of the group from the MLPS with strong interests in sanitary science and he played an influential role in both the foundation and work of the Association. Like Angus Smith, Crace-Calvert was a regular contributor to the programme of talks, including his: '*Air*: its Use, the Sources of its Deterioration, and the prejudicial Effects of Bad Air in and near Dwelling Houses, and especially in Sleeping Rooms'; '*Water* as an article of Food, and its importance as a means

---

48 Lecture programme handbills, Records of the Manchester and Salford Sanitary Association, GMCRO (Ref: M126/5/1/49, 58, 39, 41, 91,106 and 105).

49 J.K. Crellin, 'Entry for Frederick Crace-Calvert', *ODNB*.

50 Ibid.

of the preservation of Health and Life, and the prevention of Disease'; 'On the Adulteration of Food'; 'On Food'.[51]

As part of the Association's attempt to get people to take some responsibility for their own health and well-being, a series of tracts was produced for distribution across the two towns. The titles included:

1. What is Man, in respect to his Physical Constitution.
2. Hints to Working People about the Houses they Live in.
3. Facts about Health worth Recollecting.
4. Hints to Working People about Personal Cleanliness.
5. Hints to Working People about Clothing.[52]

Unfortunately, only the titles and not the detailed texts have survived in the archives.

From the surviving records of the Association it is difficult to gauge the impact of such well-intentioned activities and how they may have contributed to improving personal hygiene and health or even broader public health concerns. Although its activities changed, the Association continued until 1934:

> Committees of affiliated societies were also formed – Ladies Branch (till 1879 the Ladies Sanitary Reform Association), the Noxious Vapours Abatement Association, Committee for Securing Open Spaces for Recreation, Children's Holiday Fund and the Cheap Meals Committee. The Association continued till 1934 concerning itself with all matters relating to health and sanitation, including the provision of public parks, sanatoriums [sic] for tuberculosis sufferers, pollution of the air and rivers, the care of the mentally ill, and the use of poisons in the home.[53]

For Angus Smith, working for the Association provided a number of valuable opportunities: studying the prevailing insanitary conditions; judging how the application of science, especially chemistry, might ameliorate the worst elements; and working with others aspiring to improve public health.

---

[51]   Records of the Manchester and Salford Sanitary Association, GMCRO (Ref: M126/5/1/39, 41, and 106).

[52]   Records of the Manchester and Salford Sanitary Association, GMCRO (Ref: M126/5/1).

[53]   Introduction to the Records of the Manchester and Salford Sanitary Association, GMCRO (Ref: M126). See entry for MSAA on www.a2a..

Chapter 4

Sanitary Science, Disinfectants
and 'Acid Rain'

As soon as Robert Angus Smith arrived in Manchester in 1843 he came face to face with the appalling environmental conditions that existed there, not just the darkened skies from black smoke but also the insanitary conditions in which people were forced to live. Such experiences greatly influenced the direction of his future scientific work as he became absorbed in the factors affecting the quality of the air and in the possibility of using disinfectants to improve the insanitary conditions and control contagious diseases. Alongside his work as Inspector with the Alkali Inspectorate, the government body charged with enforcing the terms of the Alkali Act 1863, these investigations and his many associated publications were to define Angus Smith's contributions to scientific advancement and their application to public health. His membership of the Manchester Literary and Philosophical Society (MLPS) and his work for the Manchester and Salford Sanitary Association (MSSA) brought him into contact with many other similar-minded people with a desire to improve living conditions in Manchester.

Disinfection

From the 1830s and 1840s Victorians became increasingly concerned about cleanliness and dirt in response to diseases such as cholera and the publication of Chadwick's *Report on the Sanitary Condition of the Labouring Population of Great Britain.*[1] Mary Douglas has usefully expressed dirt as being 'matter out of place'.[2] While there was certainly a religious connotation to the idea of cleanliness, there was also a strong scientific interest in trying to eradicate the unhealthy living conditions that increasingly large sectors of the population experienced within large towns and cities. The ability to address these public health concerns was predicated on a detailed understanding of the causes of putrefaction and disease,

---

[1]   It is difficult to provide an accurate date, but Chadwick's report in 1842 drew attention in fine detail to the insanitary conditions that spread diseases such as cholera and typhus. The 1846 Baths and Washhouses Act was another signpost acknowledging the danger of dirt. See Katherine Ashenburg, *The Dirt on Clean: An Unsanitized History* (New York, 2007), pp. 161–97.

[2]   Mary Douglas, *Purity and Danger: an Analysis of Concepts of Pollution and Taboo* (London, 1966), p. 2.

and then adopting the appropriate actions and materials to reduce their impact. At the very same time when there was the greatest need for this understanding and application, between 1840 to 1880, there was a great deal of uncertainty about the theoretical basis of infection and contagion. There were three conflicting theories of disease causation: the miasma theory, the zymotic theory and the germ theory. All these theories had their high-profile promoters and an associated band of supporters. It was into this maelstrom of uncertainty that Angus Smith entered when he started his work on sanitary conditions associated with air and water. Studying his work over this period gives a valuable insight into the nature of the various theories, their possible compatibility and the benefits they provided in improving public health.

While the miasma theory has a history going back to Greek times, it was the theory promoted by many of the leading figures in public health in the period to the 1840s. The word *miasma* comes from the ancient Greek for pollution; the *Oxford Shorter Dictionary* defines *miasma* as 'infectious or noxious exhalations from putrescent organic matter'. It was used to describe the spread of the plague or Black Death during the Middle Ages. The theory's concern with the environment, whether air or water, concentrated attention on fumigation and deodorization.[3] Angus Smith was greatly influenced by Edwin Chadwick and his support for the miasma theory, but other influential proponents included Dr Thomas Southwood Smith, the public health reformer, and Florence Nightingale, the hospital and health reformer.

The second theory of disease causation had a distinctly chemical basis and drew heavily on the organic chemistry propagated by the German chemist, Justus Liebig, in the 1840s. This attributed diseases to fermentation products and 'particular diseases were thus caused by the distinct stages of decomposition and by the differential decomposition of different materials'.[4] Disinfection therefore required fermentation processes and associated decomposition to be terminated through the agency of disinfectant materials.[5] Angus Smith became a strong adherent to this theory because of its chemical basis that he could understand quite readily and because of his strong attachment to the ideas of his former mentor.

From the 1860s Louis Pasteur's work in advancing the germ theory of disease began to be taken seriously. Even though the Dutch microscopist Anton van Leeuwenhoeck had observed microorganisms with his crude microscope in the 1670s (including bacteria in 1676) and thereby became the father of microbiology,

---

[3]   It was not uncommon to see people wearing headdresses with a large beak in which oranges or other pleasant-smelling fruit was placed to counter the unpleasant smell of the surroundings through which the person was walking.

[4]   Rebecca Whyte, 'Changing Approaches to Disinfection in England, c.1848–1914'. PhD diss., University of Cambridge, 2012, pp. 5–6.

[5]   Justus Liebig, 'On Poisons, Contagions, and Miasmas', *Report of the Tenth Meeting of the British Association for the Advancement of Science in Glasgow, August 1840* (London, 1841), pp. 72–3.

it was not until the 1860s that Pasteur investigated the role of microorganisms in fermentation and showed they were responsible for souring beverages such as beer, wine and milk. These investigations led Pasteur to expound the germ theory of disease. Robert Koch, the German microbiologist, was then able to show that individual diseases were caused by a specific microorganism (bacterium). In 1877 Koch used his gelatine-culture technique to isolate the anthrax bacillus and in 1882 he isolated the tuberculosis bacillus, for which he was awarded the Nobel Prize in 1905.[6]

Angus Smith became aware of Pasteur's work through the scientific literature, although the full theory and therefore the impact of Pasteur's discoveries was felt only after Angus Smith's death in 1884. As we shall see from his evidence to the Royal Commission on Cattle Plague in 1866 and his subsequent articles, Angus Smith came to terms with these different theories, trying to find compromises between them because it appeared that each operated effectively in particular circumstances.[7] Certainly he was reluctant to abandon any of the theories that had served so well for so long and yet had never provided a complete understanding that would allow disease control and public health on a broad front to move forward in the serious manner that the situation in major towns such as Manchester necessitated. For Angus Smith it was work in progress, and he was fully prepared to continue his detailed studies on disinfection and disinfectants alongside his work for the Alkali Inspectorate.

The Influence of Edwin Chadwick

Robert Angus Smith was greatly influenced by Edwin Chadwick in his work on sanitary science.[8] They first became known to each other when Angus Smith assisted Lyon Playfair with his Report on the Large Towns of Lancashire as part of the Royal Commission of the Health of Towns, of which Chadwick was a commissioner, and they were then to remain in regular contact through meetings and correspondence for the rest of Angus Smith's life.[9] By the time Angus Smith became known to Chadwick, Chadwick had produced his tour de force publication in 1842, *Report on the Sanitary Condition of the Labouring Population of Great*

---

[6]   Angus Smith was aware of Koch's gelatine technique by October 1882. See Robert Angus Smith, 'Note on the Development of Living Germs in Water', *Proceedings of the MLPS*, 22 (1883), pp. 25–32.

[7]   John M. Eyler, 'The Conversion of Angus Smith: The Changing Role of Chemistry and Biology in Sanitary Science, 1850–1880', *Bulletin of the History of Medicine*, 54 (1980), pp. 216–34.

[8]   For a summary of Chadwick's role, see Christopher Hamlin, *Public Health and Social Justice in the Age of Chadwick* (Cambridge, 1998), pp. 1–15.

[9]   A collection of letters between Robert Angus Smith and Edwin Chadwick is held in the Special Collections Library, University College London.

Britain, which in many ways provided the impetus for the urgent sanitary reforms to improve public health and reduce the high mortality rates that were so prevalent, particularly in the large towns.

Chadwick's support for the miasma theory of disease had a marked influence on Angus Smith, who was left struggling to find a compromise with the chemical theory of Liebig that he found compelling and easier to accept because of its inherent chemical ideas. Chadwick and Angus Smith respected each other's opinions, exchanging their scientific papers even as draft versions. For Chadwick, Angus Smith became his preferred analyst of water samples. An interesting case in point arose from the 1848 Public Health Act, following which some 1,000 samples of water from all over the country required analysis. There were the usual logistics to consider – whether the analyses should be done in London, where the samples would be sent initially, including to Angus Smith's laboratory in Manchester, what the cost per sample would be, and the scale of the analysis.[10] As usual, the Treasury was reluctant to allocate any funding for such fees, indeed for any expenditures.

In February 1874 Chadwick sought Angus Smith's advice about the relative merits of granite setts and asphalt that were being used for the roads in the London. The initial stage was to collect from the roads the surface materials created by the wear and tear of vehicles and the ubiquitous horse dung. Not surprisingly, 'the greater proportion of the matter his machine [Whitworth's] took up was horse dung'.[11] The dilemma was that

> The roadway of Cheapside was covered three years and a half ago with two inches and a quarter of asphalt. The asphalt is reduced in thickness about half an inch, but whether by abrasion or compression and in what proportion by abrasion cannot be made out, for owing to some irregularity of surface, it does not appear to reduced [sic] in weight. But your analysis would seem to make out there 78 per cent of mineral matter against 21 per cent of organic and volatile matters.[12]

Chadwick goes on to pose a further question before ending:

> What deductions does yr analysis in yr view afford for comparison between the amount of product from wear on the smooth asphalt as compared with the product from the rough granite?[13]

---

[10]   Letter from Alex Baine (Secretary to Edwin Chadwick) to Robert Angus Smith, dated 8 January 1849. CC (Ref: Letter 48). Letter from Robert Angus Smith to Edwin Chadwick, dated 9 January 1849. CC (Ref: Letter 50).

[11]   Letter from Edwin Chadwick to Robert Angus Smith, dated 13 February 1874. CC (Ref: Letter 86).

[12]   Ibid.

[13]   Ibid.

Unfortunately, Angus Smith's response, if there was one, has not survived.

Visits abroad could result in a long letter on some interesting subject related to their work. In July 1876 Angus Smith visited Rome, presumably for a holiday. Knowing of Chadwick's increasing interest in the use of cement as a building material, Angus Smith wrote about some of the remarkable buildings and structures in Rome that used cement:

> I know that the Romans used no great masses of rock such as were used in Egypt and elsewhere in the East, or even in our own country in rude stone monuments, I associated solid strength by solid stone buildings with all the Roman structures. Whether it was the neglect of writers of books, or my neglect in reading their books I cannot tell, but when I went to Rome, I was astonished to find that as a rule the great buildings of Imperial Rome were made of rubble held together by Cement ... There are walls built of this rubble of a height which would astonish us even in stone. The rubble is made of broken stone. Rome of the age preceding is broken up into pieces and put into a frame and cemented; broken statues and broken brick, broken walls, marble, old tiles, gravel and rubbish, all cemented and made into new buildings.
>
> Rome teaches us that buildings as good as the world has may be made of cement, and may last longer than they are required by a nation. Buildings as a rule are not wanted for more than three hundred years; the habits of man change enough in that time to demand new ones.[14]

The views expressed by Angus Smith in this letter came at a time when the priority was still to improve buildings, particularly living accommodation, only just before the founding of the Society for the Protection of Ancient Buildings (1877), of which William Morris was the leading figure. Nevertheless, the letters quoted give an insight into the nature and diversity of the dialogue between two outstanding figures of nineteenth-century sanitary science.

Disinfectants, Carbolic Acid and McDougall's Powder

Soon after his arrival in Manchester, Angus Smith started investigating the nature of any organic materials found in air and in water, and as these investigations continued he became increasingly concerned about the scale of sewage, of both human and industrial origins, then contributing to the unhealthy conditions in Manchester. It was through the MLPS that Angus Smith met Alexander McDougall and Frederick Crace-Calvert, who were to be instrumental in advancing the possibility of using

---

[14]   Letter from Robert Angus Smith to Edwin Chadwick, dated 28 July 1876. CC (Ref: Letter 96).

carbolic acid in the form of a disinfectant. This would result in Angus Smith and McDougall taking out a joint patent for a disinfectant powder in 1854.

Frederick Crace-Calvert was elected a member of the MLPS in April 1847, some two years after Angus Smith. The Society's activities brought the two men together with their shared interest in chemistry and public health. Crace-Calvert had settled in Manchester following his return to England from France and it may be as the result of his experiences at the Musée d'Histoire Naturelle, and later during his work in the dyeing and cotton industry that he developed an interest in phenol (or carbolic acid, as it was more widely known at the time), which led by 1857 to a commercial process for its large-scale production.[15] Although Crace-Calvert reported his manufacturing success in 1867, the same year that Joseph Lister's work using carbolic acid as a surgical antiseptic was announced, he had explored the potential of carbolic acid for treating sewage for many years.[16]

It was probably through the MLPS that Angus Smith and McDougall first heard of the possibility of using carbolic acid to treat sewage, although their initial interest was the removal of sewage and its application as a manure to the land, given its abundant composition of organic nutrients; as we noted earlier, the awful smell had to be treated with a deodorizer. From about 1847 Angus Smith and McDougall worked on sewage manure, and as the historian Alan Gibson has pointed out, they combined magnesia (to retain the ammonia and phosphorus) with sulphur dioxide (as the sulphite) that would act as a powerful deodorant.[17] Interestingly, they had tried using carbolic acid in the same way but found it less effective, while a combination of magnesium sulphite with carbolic acid worked very well and was worth exploring commercially.[18] McDougall was already manufacturing chemicals recovered from waste products, but since Angus Smith was not interested in joining his friend and colleague in chemical manufacture it was left to McDougall to manufacture the disinfectant powder. Given the bulk nature of production for the powder, and because of the limited supply and expense of magnesia, it was decided to use a mixture of calcium magnesium carbonate (magnesian limestone). The disinfectant powder was patented as B.P. 1854/142 in 1854.[19] In treating the town's sewage, the patent specified:

> In the application of our Invention to the sewage of a town, if the object be merely the removal of the offensive smell, it will be sufficient to introduce the preparation into the sewers through the grids or other openings to the surface,

---

[15]   Phenol was used in taxidermy and in the production of picric acid and rosolic acid.

[16]   Frederick Crace-Calvert, 'Manufacture and Properties of Carbolic Acid', *The Lancet* (14 December 1867), pp. 733–4.

[17]   Gibson, 'Sanitary Science', p. 3.16.

[18]   Ibid.

[19]   B.P. 1854/142, *Deodorizing and Disinfecting Sewage Matters, and Separating Manure thereform*, registered in the names of Robert Angus Smith and Alexander McDougall.

but if it be also desired to preserve the manure, it will be necessary to provide reservoirs or receptacles (to be used alternately), in which the sewage may be allowed to stand while the matters which have been precipitated by the action of the disinfectant shall have time to subside, when the clear water standing above the deposit may be allowed to run off.[20]

The powder was marketed under the name McDougall's Disinfectant Powder, and although Angus Smith's name was often associated with the powder, he declared no financial interest after recommending the powder in evidence to the Royal Commission on the Cattle Plague. Angus Smith also got into trouble when appearing to indirectly promote the benefits of the powder in a talk to the MLPS, where promotion of any product in which the person had an interest was prohibited. The powder was found to work very efficiently by a number of independent users. One of McDougall's first trials was in the stables at the Manchester Horse Barracks of the 3rd Light Dragoons. In fact, it worked so well for this purpose that it was recommended to the Secretary of War, with the result that the powder was used on every transport ship moving horses to the Crimea.[21] The disinfectant powder was still being used for such purposes in 1884 and its continuing high-profile promotion brought it to the wide attention of other potential users, including the authorities responsible for town sewage.

During 1856 the Staffordshire town of Leek was experiencing an epidemic of fever probably due to insanitary conditions in the town. Following an approach by the medical authorities, McDougall treated the cesspools in the town that were thought to be responsible for the outbreaks of fever; the disinfectant powder was found to work very efficiently. When Angus Smith reported this important result during his lecture 'On Disinfectants' at the Royal Society of Arts in April 1857, and others in the audience also confirmed the powder's efficacy, the meeting chairman, Lyon Playfair, reminded those present that while such disinfectants were welcome there was no substitute for cleanliness in fighting insanitary conditions.[22] Leek was not the only town where McDougall applied his disinfectant powder. He worked for many years trying to treat sewage in Carlisle. Not only did the work involve treating the sewage, but the treated sewage was to be used as a manure, as in Angus Smith and McDougall's original plans. As Alan Gibson summarized:

> The sewage from about 22,000 of Carlisle's 30,000 inhabitants was used in this way. It was treated by the addition of 12 gallons [55 litres] of disinfectant per day to about 500,000 gallons [2,273,000 litres] of sewage, at a cost of £25 per annum. Presumably the addition of the powder to the sewage did not prove satisfactory, and in 1859 McDougall took out another patent, without [Angus]

---

20   Ibid., p. 4.

21   Gibson, 'Sanitary Science', p. 3.20.

22   Robert Angus Smith, 'On Disinfectants', *Journal of the Society of Arts*, 5 (1857), p. 341.

Smith, for the preparation of a liquid disinfectant, made by treating heavy oil of
tar with alkali, with 3% nitric acid added to help solubility and sometimes 2%
of a metallic salt.[23]

This disinfectant seems to have worked much more effectively and the sewage
manure was applied to farmland for grazing cattle.

Following the Great Stink in London in the summer of 1858, when the
overwhelming smell from the heavily sewage-laden River Thames caused
Parliament to consider moving to Hampton Court, much attention was focused
on a means to provide a lasting resolution to the problem of sewage and its
treatment.[24] However, the committees set up to make recommendations floundered.
Angus Smith and McDougall offered a solution based on their disinfectant
powder, alongside many other proposals. One report in 1858 even included the
recommendation:

> I believe that it has been proved in several of our cavalry barracks and by
> dairymen, that the best disinfectant for stables, cowsheds and piggeries, at the
> same time greatly improving the health of the cattle where used, is a powder
> manufactured by Mr. McDougall of Manchester.[25]

The powder was also used for the Tottenham Court Road sewer, and again the
results demonstrated its efficacy. Unfortunately such trial results and endorsements
did not lead to a contract, and a later report by Wilhelm Hofmann and Edward
Frankland mentioned the McDougall powder only alongside other possible
disinfectants.

Meanwhile McDougall continued with his successful work on sewage in
Carlisle, which was beginning to attract a good deal of attention, in the hope of
applying the disinfectant powder to treat sewage in many other towns in Britain.
As it turned out, the Carlisle work was to attract the attention of Joseph Lister and
influence his pioneering treatment of surgical wounds with antiseptics. Lister had
initially taken an arts degree at University College London and, having observed
the first surgical operation with ether in Europe performed by Robert Liston on
21 December 1846, decided to follow a medical career, first at Edinburgh and
then as Chair of Surgery at Glasgow University.[26] Lister had become aware of
Pasteur's germ theory of disease and began to understand why wounds exposed to

---

[23]   Gibson, 'Sanitary Science', pp. 3.23 and 3.24.

[24]   The final resolution of the Great Stink came in 1859 when the sewer system
proposed by Joseph Bazalgette was approved. It serves the City of London to this day.

[25]   T. Baker, 'On the Plan Suggested by the Government Commissioners for Disposing
of the Metropolitan Sewage', *Journal of Society of Arts*, 6 (1858), p. 418.

[26]   Christopher Lawrence, 'Entry for Joseph Lister', *ODNB*. The first operation using
ether anaesthesia was performed by William Morton at Massachusetts General Hospital on
30 September 1846.

air (and any microorganisms or germs in the air) were likely to undergo infection and suppuration. But how could this suppuration be treated and prevented? Pasteur's experimental work suggested filtration, heat or the application of suitable chemicals, and while the first two were not practicable, Lister remembered the remarkable manner in which carbolic acid had been used to treat the sewage in Carlisle. A colleague supplied Lister with a less than pure form of carbolic acid for his investigations, but it soon became apparent that the most efficient carbolic acid was the pure form produced by Frederick Crace-Calvert. Lister's subsequent application of carbolic acid as a spray on the surgical wound was to provide a remarkable reduction in gangrene, amputations and deaths resulting from septicaemia. For Angus Smith and McDougall, their work brought health benefits that were completely different from those they were expecting or had planned.

By the mid-1860s, having achieved an even higher public profile following his election as a Fellow of the Royal Society (1857) and his appointment as Inspector of the Alkali Inspectorate (1864), Angus Smith was in correspondence with Florence Nightingale, the hospital and public health reformer. She was eager to learn about the latest advances in public health in Britain as part of her work to improve sanitary conditions in India. In 1865 she wrote to Angus Smith congratulating him on his first annual report as Inspector of the Alkali Inspectorate; this 'demonstrated not only her interest in scientific advance but her enormous confidence in the power of science and her assiduous reading even in areas outside her field'.[27] Nightingale was keen to have the best available method of analysing water for application by the authorities in India and, while satisfied that the best method was the one adopted by Angus Smith, she found working with him very arduous and became rather impatient at times.

> I think Dr Angus Smith is as difficult to manage as the whole India government. He writes one thing, then he writes the reverse, then he listens to what his 'nephews and nieces in Argyleshire' [sic] say and tears up the paper. (But I have it in type.) And there is now scarcely a single word in this, the sixth revise, of what there was in the first. But he is the only man in Europe who can do it. And this is well worth all the trouble. When it has reached the sixtieth revise, I shall make the India and War Offices circulate it.[28]

In the end she had 600 copies printed and distributed at her own expense.

[27] Lynn McDonald (ed.), *Florence Nightingale on Society, Philosophy, Science, Education and Literature* (Waterloo, Ontario, 2003), p. 652. See also Letter from Florence Nightingale to Robert Angus Smith, 15 June 1865 (Private Collection of Hugh Small, copy Balliol College, Oxford).

[28] Letter from Florence Nightingale to Lord Stanley (Chair of The Royal Commission on the Sanitary State of the Army in India), dated 22 October 1865. LRO (Derby Collection: Ref: 920/15/88). See Gérard Vallée (ed.), *Florence Nightingale on Health in India* (Waterloo, Ontario, 2006), p. 534.

However, the matter was not resolved at this time, for in a letter dated 12 May 1866, Angus Smith appears to query the legitimacy of his method of analysis in all circumstances and claimed that further work was necessary. But matters moved forward only slowly. Six weeks later Angus Smith informed Florence that he must take a long-planned holiday immediately, and so work on the paper about water analysis must wait. Even more annoying for Florence, Angus Smith started discussing the fee required, to which Florence responded that the paper must be quite short 'to enable it to be useful' and 'if you cannot tell us the fee then you will have to take the chance of one getting something'.[29] Florence's impatience was probably due to the long illness she had suffered the previous winter, and her determination to overcome the unnecessary officialdom and petty bureaucracy she found around her. It is not known whether Angus Smith ever completed this work for Florence Nightingale.

Cattle Plague

In June 1865 an outbreak of cattle plague, or rinderpest, as it was known in veterinary circles, was reported at Islington cattle market in London. It was endemic on mainland Europe but this was the first reported occurrence in the UK since the long outbreak between 1745 and 1757.[30] The disease primarily attacked cattle, although other animals could catch it, and was transmitted by breath or by infected drinking water. It was highly contagious and the outbreak in 1865 spread very quickly through the rest of the country, via East Anglia and then on to Cheshire and Yorkshire, where there were large dairy herds. By September 1865, 13,000 cattle had been affected and by January 1866 the number of animals infected had grown to more than 120,000 with no end in sight to the spread. Nevertheless, by November 1866 the disease had to all intents and purposes vanished.

What is especially interesting about this outbreak of rinderpest is the reaction of officialdom to the disease and in particular how actions to contain it were reviewed at a time when theories on infection and the use of disinfectants were still subject to considerable speculation, even among experts. Cattle movements were the principal cause of the spread of the disease, aided by development of the railways, and in July 1865, under the direction of John Simon, the Privy Council's Medical Officer, a Parliamentary Order in Council was approved restricting cattle movements and requiring all affected cattle to be slaughtered and quarantined. But because the farmers and slaughter-houses remained lax in their actions, the disease did not abate. There was much debate and discussion on possible ways of treating the disease and, after what must have appeared an interminable period, a Royal

---

[29]   Letter from Robert Angus Smith to Florence Nightingale dated 22 June 1866 and a pencil note by Florence Nightingale, dated May 1866. BL (Ref: Add. 45799).

[30]   It was only on 8 August 2011 that the United Nations declared rinderpest eradicated, only the second disease after smallpox.

Commission was set up in September with the Earl of Spencer as Chairman. The Commission reported on three occasions: 31 October 1865, 5 February 1866 and 1 May 1866. The first report gave details of the origin, nature and diagnosis of the disease, with a short summary for each county and a map indicating its spread, while the second concentrated on the county-by-county summary with the maps. It had become clear that the Order in Council on cattle movements and slaughter had not worked (or more likely the Order was being widely flouted) and that there was much disagreement among the commissioners about what additional measures to put in place. It was only the third report that made any serious attempt to address the spread of the disease on a scientific basis, and this was achieved through a series of recommendations by various experts, including Angus Smith, William Crookes (editor of *Chemical News*), George Varnell (Professor of Anatomy with Physiology at the Royal Veterinary College, London) and Lionel Beale (Professor of Physiology and of General Morbid Anatomy at King's College London).[31]

By the mid-1860s Angus Smith was seen as an expert in the field of disinfection and disinfectants, and therefore it was not too surprising that the Commission asked him to undertake some experiments as an aid to their deliberations. As if to confirm the Commission's continued uncertainty over its progress, the only advice Angus Smith received in this connection came from Lyon Playfair, one of the commissioners and his former mentor, who suggested that he 'make the report as practical as possible'.[32] Angus Smith used the opportunity to provide a short history of the subject and outline his own philosophy based on Liebig's chemical theory, while also recognizing the emerging germ theory of Pasteur. His paper then went on to give an account of the current practice of disinfection, including details of McDougall's Disinfecting Powder before providing an extensive summary of the disinfectant powers of a large number of chemicals in different situations, whether preserving meat, disinfecting water, disinfecting air, the action of ozone, the treatment of dead cattle to be preserved for manure or disinfection to prevent cattle plague. The report included controlled experiments on the effects of a range of disinfectants in different situations over varying periods of time. Angus Smith might have continued his report with even more experimental examples but the Commission's report was due for publication. In 1869 Angus Smith took the opportunity to publish a fuller summary of his work as *Disinfectants and Disinfection*. In his report and later book he acknowledged William Crookes's experiments; the men had a great deal of respect for each other's scientific work When Crookes reviewed *Disinfectants and Disinfection* in *Chemical News* he was

---

[31]   For the contribution of William Crookes to the Commission, see Brock, *Crookes*, pp. 90–101.

[32]   Robert Angus Smith, 'On Disinfection and Disinfectants', Report to the Cattle Plague Commission, 25 April 1866, in *Third Report of the Commissioners appointed to enquire into the origin and nature, etc., of the cattle plague, and an appendix*, P.P. 1866 (3656), pp. 155–86.

magnanimous in his comments – 'by common consent Dr Smith has become the first authority in Europe on the subject of disinfection'.[33]

Air Quality Investigations

Ever since being confronted by the dense black clouds hanging over Manchester as he walked into town one summer's day in 1844, Angus Smith had taken a keen interest in what the air contained and how its variable content might affect health and well-being. He took every opportunity to analyse the air in a wide range of locations – in different towns, in the countryside, at sea and at the seaside, at different altitudes, in different social and industrial settings, and in mines; and how the variation in content came about. Techniques were developed for collecting and analysing air samples, as well as conducting controlled experiments to test how the human body reacted to variations in air content. His investigations embraced the role of air as a carrier of disease as well as a disinfectant. While Angus Smith's work has been seen as just the accumulation of a vast body of experimental results, he was also very keen to apply this knowledge for the benefit of all members of society and in particular to improve sanitary conditions in the main urban areas. His work as Inspector of the Alkali Inspectorate, regulating the harmful vapours from industrial chemical processes, can be interpreted as just one part of his overall commitment to improving air quality.[34] In 1879 Angus Smith brought this body of work together with the publication of the book, *Air and Rain: Beginnings of a Chemical Climatology*.[35]

An interesting way to plot Angus Smith's progress on his research and investigations into air quality is to consider the chronology of his published reports. The first was 'Some remarks on the air and water of towns' in 1845 and the last was 'Detection of fire-damp' in 1879.[36] Between these two papers and over a period of almost 35 years, Angus Smith tackled a wide range of issues, including the composition of the atmosphere, the air of towns, the air in houses and workshops, fogs in Iceland, air in the mid-Atlantic, air in the London law courts, organic matter in the air, the physiological effect of carbonic acid, ventilation, absorption of gases by charcoal, distribution of ammonia, and minimetric analysis. While these papers give the broad range of topics, in order to appreciate the full extent of his investigations it is necessary to consider the contents of his report for

---

[33]   *Chemical News*, 9 (1869), p. 105.

[34]   Eville Gorham, 'Robert Angus Smith, F.R.S., and "Chemical Climatology"', *Notes and Records of the Royal Society of London*, 36/2 (1982), pp. 267–72.

[35]   Robert Angus Smith, *Air and Rain: Beginnings of a Chemical Climatology* (London, 1879).

[36]   Robert Angus Smith, 'Some Remarks on the Air and Water of Towns', *Chemical Society Memoir*, III (1845–48), pp. 311–19, and Robert Angus Smith, 'Detection of Fire-damp', *Chemical News*, 39 (1879), pp. 267–8.

the Royal Commission on Mines in 1864 and his book *Air and Rain: Beginnings of a Chemical Climatology.*

Angus Smith was one of three experts asked to report on the air of mines to the Royal Commission on Mines in 1864 in an extensive review of mining operations in Britain.[37] The air in mines was of particular concern, not just because of the occurrence of fire-damp, but also in relation to levels of oxygen and carbon dioxide, given the limited ventilation in many mines. Angus Smith was asked to consider whether any other gases or materials in the air should be monitored, such as powder smoke and organic matter. As with his report to the Cattle Plague Commission, Angus Smith took the opportunity to summarize what was known in a much longer brief, even comparing the situation with that found in the mines of Germany, as the contents of his report show (see Figure 4.1).

REPORT ON THE AIR OF MINES
By Dr. R. Angus Smith, F.R.S.

CONTENTS

Composition of the Atmosphere
  Amount of oxygen in pure air
  Analyses made by various methods
  Air deviating from the adopted standard
  Air of impure places
  Oxygen in the air in wet and in dry foggy weather
  In dwelling rooms, etc.
  In cowhouses and stables
Carbonic acid of the atmosphere
  Air of Madrid
  Hospitals
  Carbonic acid of the air in England
  Carbonic acid just outside Manchester
  Close places
  Carbonic acid in the air of Manchester
  Carbonic acid on the River Thames
  Carbonic acid in open places of London
  Carbonic acid in the streets of London
  Carbonic acid in close places in London
Analyses of the air of mines
  Method of analyses
  Measure of accuracy of the analyses
  Tables of analyses
  Summary of the analyses
General impurities of the air of mines and calculation
  of extent
Solid impurities in the air of mines
  Amount of crystals observed compared with the
    oxygen or its deficiency
Organic matter of the air

The organic matter in the mines
  The sense or smell and organic matter
Tests of carbonic acid and of ventilation
  New mode of using the baryta and lime water test
  Manganates and ferrates as tests for carbonic acid
Air of confined places
  Bad air and the sensations
  Experiments in the lead chamber
Action of the pulse
Combustion of candles
  Weight of candle burnt in air of various qualities
  Effect of moisture
Carbonic acid and ventilation
  Washing the air
  Ventilation
Heat
Gun cotton
  Combustibility of gun cotton
  Gases of combustion
Summary of information derived from Germany and German
  sources
  Hausmann on the Hartz miners
  Dr. Brockmann's work on the Hartz miners
  Abridgement of Dr. Brockmann's summary
  Remarks on the German system
  Comparison of some of the German and English habits in
    mining districts
  Historical mining
  Notes from books
  Notes on climbing

Figure 4.1   Contents of Angus Smith's report for the Royal Commission on Mines

---

[37]   The other experts were Dr A.S. Taylor (Guy's Hospital) and Dr Bernays (Professor of Chemistry, St Thomas's Hospital Medical College and physiological chemist to the hospital), who had written widely on ventilation.

Angus Smith analysed no fewer than 328 samples of air from mines in different part of England and Wales, where necessary descending into the mines to take samples. He found significant variation in the amount of oxygen and used a simple classification system, with air containing at least 20.9 per cent of oxygen considered normal; less than this proportion was impure and less than 20.6 exceedingly bad. His 328 samples divided as:

> 35 (10.6%) of the samples, normal or nearly so
> 81 (24.7%) of the samples, impure
> 212 (64.6%) of samples, exceedingly bad.[38]

In a further observation, he noted that for the mines in Cornwall and Devon, 142 were found to have defective air, with only 17 having the normal proportion of oxygen, while in Wales and Shropshire, ventilation was found to be even more defective, with 45 samples out of a total of 59 classified as exceedingly bad.[39]

When Angus Smith gave evidence to the Commission on 17 June 1863 his full report was yet to be completed but he started by summarizing his results from the analysis of air to date and explaining how he had classified the samples of air into normal, defective and bad. He was then asked about products introduced into the air when gunpowder was exploded, to which he replied:

> carbonic oxide, hydrogen, and sulphuretted hydrogen; these three as gases. Then there are sulphate of potash, carbonate of potash, hyposulphite of potash, sulphide of potassium, sulpho-cyanide of potassium, nitrate of potash, carbon, sulphur, carbonate of ammonia, sulphide of ammonium, organic matter, sand and in some cases a little arsenious acid. These are introduced into the air.[40]

The arsenious acid was formed from any arsenious ores present in the seams. He also highlighted the analytical determination of organic matter in the air using potassium permanganate solution rather than the earlier method using ammonia, since ammonia was often present in the air being tested.

When giving his evidence to the Commission, Angus Smith stated his intention of conducting some experiments in a special chamber that he hoped to get access to for a short time. Unfortunately he was subsequently unable to secure use of the chamber, so he decided to make a small room of lead in which to conduct the experiments, as his report details:

> Accordingly a chamber was made of lead, six feet long, nearly four feet unequally broad, and eight high; the cubic contents were 170 feet – it is unnecessary to describe the irregularities of the shape partly caused by the door, the windows,

---

[38]   *Report of Royal Commission of Mines*, P.P. 1864 (3389), XXIV, p. xvi.
[39]   Ibid., p. xxxvi.
[40]   Ibid., p. 458.

and the position. The whole lead of this chamber was made into one piece by having the edges melted together by the hydrogen blowpipe, as is now done in making chambers for sulphuric acid.[41]

But there had to be provision for easy escape should the need arise:

> As it was important that no one should be left in the place without having the power of escape, it was determined that there should be sufficiently large windows which in any emergency could be broken through; were it otherwise, the mere idea of being confined and at the mercy of the treacherous memory of human beings would be enough to cause much discomfort and render the experiments void.[42]

It was in this chamber that the series of experiments was carried out.

With the variable levels of ventilation in mines, concern was expressed about the levels of carbon dioxide and how they might affect a miner's ability to work over long periods. It was recognized that while other gases present in the mine shaft would affect miners, the best chemical test for air quality was the one for carbon dioxide. The method then available used either limewater or baryta, but if such a test were to be readily replicable underground and used by miners or others, it would need modifying. Angus Smith proposed a method that was to all intents and purposes the one developed by Pettenkofer in 1858.[43] The procedure was aided by a simple form of apparatus made by J.B. Dancer, the Manchester instrument maker and friend of Angus Smith's at the MLPS. This apparatus became part of the minimetric method of analysing gases and was used extensively by the inspection team of the Alkali Inspectorate. The Commissioners were also interested in whether there was a means of removing the carbon dioxide from the mine shafts; Angus Smith recommended the simple apparatus designed by Sir Goldsworthy Gurney for washing air that was subsequently installed at the House of Commons.[44]

The normal test to indicate the presence of satisfactory air in a mine was to burn a candle, and the Commissioners pressed Angus Smith about the reliability of the test. In his evidence Angus Smith was somewhat reserved, but with time to carry out his laboratory tests under varying conditions – carbon dioxide, organic matter and explosive smoke – he became convinced, and confirmed the value of this widely adopted practice:

---

[41]   Ibid., p. 241.

[42]   Ibid.

[43]   A. Pettenkofer, 'Volumetric Estimation of Atmospheric Carbonic Acid', *Quarterly Journal of the Chemical Society*, 10 (1858): pp. 292–7.

[44]   Robert Angus Smith, 'Report on the Air of Mines', in *Report of Royal Commission of Mines*, P.P. 1864 (3389), XXIV, p. 259.

A belief so long and so well established as that concerning the infallibility of the candle is not to be supposed to be entirely unfounded. It will be seen from the experiments given that, within bounds, it is remarkably well founded, and by the aid of fine measures or balances may be carried out to an extent far exceeding that which it has hitherto reached. This great value it certainly has, that it indicates impurity of air of almost every kind usual in mines, little heeding whether that impurity be pure vapours of water or poisonous carbonic acid; and to a great extent the effect of the two substances on the candle very much resembles their effect on human life or at least comfort.[45]

Interestingly, there was discussion about the validity of smell as a means of detecting the presence of stale air. Smell is the not the chemist's first choice when it comes to analytical techniques, although it is extremely valuable, even in detecting low concentrations of, for example, phenol in water. Testing by smell appeared to be common practice in the mining industry. The smell of apples or presence of an aroma was the standard test, and Angus Smith 'was extremely surprised to find in several German books that the apple smell is an ancient attribute of badly ventilated places'.[46] Presumably this test was not reliable for those miners suffering with a cold!

As in his report to the Commission on Cattle Plague and in his book, *Disinfectants and Disinfection*, much of the evidence contained in his report to the Royal Commission on Mines was incorporated into his later book, *Air and Rain: Beginnings of a Chemical Climatology*.

'Acid Rain'

Much of Angus Smith's work investigating the quality of air concentrated on the presence of carbon dioxide and organic matter, but he became increasingly concerned about the quantity of sulphuric acid and its damaging effect on the natural and human environments. At the Twenty-First Meeting of the British Association for the Advancement of Science held in Ipswich in 1851, Angus Smith presented a paper 'On Sulphuric Acid in the Air and Water of Towns' that gave a short summary of his investigations to date and stressed the need for continued research.[47] Later, in 1859, in his paper 'On the Air of Towns', Angus Smith was able to elaborate further by drawing attention to the sulphur in coal as the agent

---

[45]   Ibid., p. 254.

[46]   Ibid., p. 236.

[47]   Robert Angus Smith, 'On Sulphuric Acid in the Air and Water of Towns', *Report of the Twenty First Meeting of the British Association for the Advancement of Science held in Ipswich July 1851* (London, 1851), p. 52.

responsible for sulphuric acid in the air.[48] With Manchester's heavy industrial activities, the best available data suggested that about 4 million tons of coal were burnt there annually. He began his investigations by analysing 71 specimens of coal, and found that some had as much as 5 per cent or even 6 per cent of sulphur, although the average for specimens used in Manchester was about 1.4 per cent.[49]

The next stage of his investigation was to measure the amount of sulphurous and sulphuric acids by passing air samples through lead acetate solution using Liebig's potash apparatus.[50] The amount of sulphur acids measured varied with the air conditions, whether it was raining, and the location. As would be expected, samples collected close to chimneys contained much higher quantities. Nevertheless, it appeared that not all the sulphur was converted into sulphur acids due to the inefficient burning of coal in most hearths, whether industrial or domestic. This presented a dilemma. If the coal were burnt more efficiently, as many environmental scientists were promoting to ensure that the coal was completely converted into carbon dioxide without releasing soot and tarry residues, then the quantities of sulphur acids would be maximized.[51] The only solution might be to control the cost of coal so that those with the lowest sulphur content were cheapest but the government and local authorities were unlikely to take such direct action on prices.

Angus Smith's attention then turned to the action of the sulphur acids on stonework, bricks and mortar. He had observed how

> the stones and bricks of buildings, especially under projecting parts, crumble more readily in large towns where much coal is burnt than elsewhere. Although this is not sufficient to prove an evil of the highest magnitude, it is still worthy of observation, first as a fact, and next as affecting the value of property. I was led to attribute this effect to the slow but constant action of the *acid rain*. (Italics added)[52]

Acid rain also affected the galvanized iron roofs that were widely used, and 'it will be observed that this style of roofing is preserved in exact proportion to its distance from manufacturing districts'.[53] It is interesting to note that while his 1859 article includes the first use of the phrase 'acid rain', Angus Smith had, in his 1852 article 'On the Air and Rain of Manchester', found 'that not all rain is acid', so it was a phrase developed over time.[54]

---

[48] Robert Angus Smith, 'On the Air of Towns', *Quarterly Journal of the Chemical Society*, 10 (1859), pp. 192–235.

[49] Ibid., pp. 205–6.

[50] Ibid., p. 206.

[51] Ibid., p. 208.

[52] Ibid., p. 232. This is the first use of the phrase 'acid rain' and is cited as such in the *Oxford English Dictionary*.

[53] Ibid., p. 233.

[54] Robert Angus Smith, 'On the Air and Water of Manchester', *Memoirs of the MLPS*, 10 (2nd series) (1852), pp. 207–17.

Angus Smith will have been aware from the Peter Spence court case in 1857 that sulphuric acid in the air can severely damage trees and hedges. In his evidence to the Royal Commission on Noxious Vapours in 1876, Angus Smith estimated the annual consumption of coal in Britain at about 100 million tons, and assuming a sulphur content of 1 per cent, this would result in about 1 million tons of sulphur being converted into sulphuric acid and released into the air. While his paper of 1859 provides the first use of the phrase 'acid rain', the phrase came to have much larger significance in the 1980s when acid rain wreaked environmental havoc across continents. This is dealt with in more detail in Chapter 8.

Chapter 5

Nuisance Vapours, 'The Monster Nuisance of All' and their Legal Challenge

Airborne emissions from alkali manufacture were 'the monster nuisance of all'.[1]

The first half of the nineteenth century saw a dramatic transformation in chemical industry, with the adoption of new processes as chemical understanding advanced the shift from natural sources of chemicals to synthetic production and the dramatic increase in levels of production to meet the demands generated by mass industrialization. Many of these chemical works were located in the centre of major towns and therefore close to residential areas. Although chemical technology had advanced, aspects of process control remained rudimentary, leaving manufacturers reluctant to reuse any waste products for which there were not immediate and ready uses. Manufacturers were left to dispose of these waste products by releasing gases into the atmosphere (usually from ever-taller chimneys), by flushing liquid wastes into adjoining streams, canals and rivers, or by dumping solid waste on land surrounding the works or at sea. These waste products were frequently smelly and injurious to health, and added to the rich effluvia already present due to surrounding manufactories. It was not long after these large chemical works started production that local inhabitants began to complain, often in their local newspaper, that they caused a nuisance.

One of the main culprits of these nuisances was alkali works using the Leblanc process for the manufacture of soda. Until the 1850s copious quantities of hydrogen chloride gas were released into the atmosphere and until the 1880s large amounts of sulphur waste (also known as 'galligu' or vat waste) were dumped on land or at sea, causing damage to the environment and to people living nearby. Often these works were grouped close together, thereby adding to their impact. Successful prosecutions in courts of law either for financial compensation or to close down the works proved difficult to achieve because of arguments about the cause of the alleged damage and about pinning responsibility on one particular works where they were clustered together. It was only when the landed gentry and the wealthy landowners found their property values under attack that members of the House of Lords, many of whom had a vested interest, took action and in 1862 set up the Select Committee on Injury from Noxious Vapours. This led in the following year to parliamentary approval of the Alkali Act, the first legislation

---

[1]   Lyon Playfair, *Report of the Select Committee on Injury from Noxious Vapours*, P.P. 1862 (486), xiv, p. 99.

to control pollution from chemical works.[2] While setting a regulatory framework establishing the Alkali Inspectorate, the Act also made provision for an inspector to head the new body, and in February 1864 Robert Angus Smith was appointed the first Inspector, as we saw in previous chapters.

Angus Smith had taken an interest in air quality while living in Manchester from the early 1840s and working there as consulting chemist. His initial concern focused on blight of black smoke from burning of coal, but he was aware of the pollution from alkali works. However, his first direct experience of a pollution case came in 1855 when he was approached by Peter Spence, an alum manufacturer at Pendleton (near Manchester) and a fellow member of the Manchester Literary and Philosophical Society (MLPS), to review the operation of the works because local residents were accusing him (and the works) of causing a nuisance. Spence feared a prosecution even though he was fastidious about the efficacy of his works, and even though he sought the advice of several consultant chemists, including Angus Smith, an indictment was issued in 1857 and Angus Smith gave evidence for the defence when the case came to court. The conduct of the court case and its outcome were to have a dramatic, perhaps even traumatizing, effect on Angus Smith and probably influenced his leadership of the Alkali Inspectorate as it emerged during the 1860s, 1870s and 1880s. By reviewing the background to Spence's alum works, and the events leading up to the court case and the court proceedings, it is possible to appreciate why successful prosecutions proved difficult and how expert witnesses (and their evidence) might be used to manipulate the verdict.

Peter Spence and the Pendleton Alum Works

Alum had been an important chemical from medieval times, and was used mainly as a mordant, but later also in printing, medicine, sewage treatment and in paper manufacture. Unfortunately, a less worthy use was to adulterate flour, and such flour gave bread some strange and undesirable characteristics as well as harming the consumer.[3] Whitby, on the north Yorkshire coast, was the location of many alum works, making use of the natural resources of alum there,[4] but a major manufactory was established at Hurlet (Renfrewshire), which made use of exhausted coal shales.[5] Chemically, alum is the name given to a specific chemical,

---

[2]   For earlier attempts on pollution legislation, see Ann Beck, 'Some Aspects of the History of Anti-Pollution Legislation in England, 1819–1954', *Journal of the History of Medicine*, 14 (1959), pp. 475–89.

[3]   J. Sheridan Muspratt, Chemistry, Theoretical, Practical and Analytical as Applied and Related to Arts and Manufactures (2 vols, Glasgow, 1860), vol. 1, pp. 149–76.

[4]   Works at Boulby provide an interesting example of the north Yorkshire alum industry. See Kevin Quinn, *Boulby Alum: The Works Diary of George Dodds 1772–1788*. Research Report Number 9 of the Cleveland Industrial Archaeology Society (2011).

[5]   Archibald and Nan L. Clow, *The Chemical Revolution* (London, 1952), p. 238.

potassium aluminium sulphate, which occurs naturally, as well as to a group of chemical compounds with a related aluminium sulphate composition.[6] Peter Spence was born in Brechin in 1806, and it was while working at a gasworks in Dundee that he developed his interest and knowledge of chemistry.[7] But his important chemical breakthrough came later during his time at Burgh-by-Sands, near Carlisle, when he discovered a method for making ammonium alum from coal shale found nearby and the ammoniacal liquor from gasworks. Spence moved to Manchester in 1846 and in partnership with Henry Dixon built the large alum works at Pendleton. It was here that Spence also built large lead chambers for manufacturing sulphuric acid, another essential chemical in his manufacture of alum.[8]

Local residents, sensing an unpleasant odour in the air, looked for the likeliest culprit nearby and their attention was drawn almost like a magnet to the recently built chemical works with their multiple tall chimneys. But other manufactories, such as tanneries, gasworks, breweries and glue works, made their own contribution to the odour in the air. For Spence, the earliest accusations were attributed to the carts transporting the ammoniacal liquor from the gasworks to his alum works,[9] and once the protests had started, a momentum built up, as the stream of letters to the *Manchester Guardian* shows.[10] Anxious to avoid offence, Spence tried to appease the accusers by entering into correspondence or by meeting them at the works, but his efforts were to no avail as the letters continued. Some of the letters sought to draw the attention of Salford Corporation to the accusations, demanding legal action to stop the nuisances identified as ammonia, hydrogen sulphide and sulphur dioxide,[11] or removing the source of the nuisance completely by closing the works, and the Salford Nuisance Inspector was drawn into investigating the accusations.

Indictment against Peter Spence

At the time of the early accusations in 1854 Spence had asked Angus Smith, as a consulting chemist and a fellow member of the MLPS, to inspect the Pendleton works and offer advice on how best to prevent any nuisance gases escaping. Angus Smith found few gas leaks and, while offering minor modifications to the plant and its operation, he was convinced that the unpleasant odours referred to in the

---

[6]   The chemical formula is $KAl(SO_4)_2.12H_2O$, while the structure is $[K(H_2O)_6]$ $[Al(H_2O)_6].(SO_4)_2$.

[7]   Frank Greenaway, 'Entry for Peter Spence', *ODNB*.

[8]   Charles Singer, *The Earliest Chemical Industry* (London, 1948), p. 283. The partnership with Dixon ended in November 1856.

[9]   Correspondence, *Manchester Guardian*, 21 April 1856. See also *Transcript of Diary of Peter Spence*, p. 38. Catalyst Science Discovery Centre.

[10]   Correspondence, *Manchester Guardian*, 25 and 29 April 1856.

[11]   Correspondence, *Manchester Guardian*, 26 April 1856.

public accusations were not caused by the alum works but were more probably coming from manufactories surrounding the works.

By October 1854 the landlord of the Pendleton site was getting anxious as a result of the increasingly aggressive tone of the letters to the newspapers, and instructed his solicitors to indict Spence and the works for causing a nuisance. The solicitors, Messrs Sale, Worthington and Sons, appointed Angus Smith and Frederick Crace-Calvert, Professor of Chemistry at the Royal Manchester Institution and a fellow member of the MLPS, to obtain scientific evidence to support a prosecution. Spence, rather than deny access or obstruct, 'Threw the works open to them, authorised them to come at any time they chose, remain as long as they found convenient, and that all the operations would produce as usual'.[12]

Angus Smith and Crace-Calvert visited the works on 7, 11 and 18 October, and noted while approaching the works a mixed odour in the air that they attributed to the various manufactories in the vicinity. Their findings identified few defects in the plant or its operation, but noted the escape from a chimney of some sulphurous acid gas that was neutralized with ammonia; the amount of acid gas in the flue was 0.147 grains per cubic foot;[13] there was some leaking of sulphurous acid between sulphur stove and lead chamber. With some simple modifications these leaks could be remedied. Angus Smith and Crace-Calvert took the opportunity to inspect the surrounding neighbourhood and could find no injury to the vegetation. The report's recommendations were: more condensers; flues made less pervious; an improved method of feeding the sulphur ovens; and some improvements to prevent escape of hydrogen sulphide. While these amounted to small-scale modifications rather than major reconstructions, the report would have reminded Spence of the important role of the site manager or foreman in operating the individual processes effectively at all times to prevent gas leaks. Nevertheless, the general standing of the works was expressed rather eloquently in the final part of the report:

> their [Spence and Dixon's] works are most extensively known, both at home and on the continent, as a model of scientific and economical arrangement; that they have been seen by nearly all the great continental chemists who have visited England, and also by the most noted chemical professors in this country, – to all of whom the utmost freedom of inspection has been accorded, and they are aware that their manufactory has been spoken of repeatedly, by some of the eminent foreign chemical professors, in their public lectures, in the high style of commendation.[14]

---

[12]   *Observations in refutation of certain charges preferred against the Alum Works of Messrs. Spence & Dixon ... With an examination of Dr. Smith and Profr. Calvert's Report on the same [and copy of that Report]*. (Manchester, 1855), p. 5.

[13]   A grain is 64.79891 milligrams. The level of accuracy expressed is unrealistic given the measuring ability at the time.

[14]   *Observations in refutation of Alum works*, p. 10.

The report's final comment referred to the lack of incriminating evidence found by the Salford Nuisance Inspector.

The report confirmed Angus Smith and Crace-Calvert's overriding view that the works was generally well run. With the report's practical recommendations, should Spence and Dixon agree to implement them fully, a prosecution could be avoided. A copy of the report was subsequently passed to Spence and Dixon by the solicitor, who told them that if they agreed to implement the recommendations fully then he would persuade the landowner not to prosecute.

While there is no evidence that the recommendations were put into effect, it is reasonable to assume that they were, given Spence's stated intentions in managing the works in a responsible manner. But the accusations continued unabated, as in so many other nuisance cases. Spence probably consulted Angus Smith again for reassurance but, feeling the need of further independent opinion, turned to Edward Frankland, Professor of Chemistry at Owens College, Manchester. Frankland seemed to confirm the findings of Angus Smith and Crace-Calvert, concluding that if the processes were carefully controlled no nuisance gases would escape. Spence must have hoped that, with this further inspection, his concerns about nuisances and having to close the works were at an end, but these hopes were dashed when an indictment for causing a nuisance was issued in early 1857.[15] However, if this was not enough to contend with, Spence had to bear another blow, for, with the court case being heard in a criminal court rather than a civil court, Spence would be prevented from giving evidence. Given the importance of the works to the alum trade and the sustained campaign of accusations over many years, it is not surprising that the case attracted a good deal of press and public interest. While the outcome was to have serious consequences for Spence and his business, it also had a far-reaching influence on the conduct of Angus Smith's future career and on the role of expert witnesses in court cases.

Court Case Proceedings

The case was heard before Mr Baron Channell in the summer assizes of the northern circuit in Liverpool on 21 August 1857, with the prosecution led by Mr Wilde QC and the defence by Sir Frederick Thesiger QC.[16] The proceedings of the case followed those of other nuisance cases, with both parties calling a variety of witnesses to give evidence, including many local residents claiming to have experience of the alleged nuisance, followed by expert witnesses such as consulting chemists, land agents, local government inspectors or medical practitioners, chosen for their professional standing and ability to add gravitas to the proceedings and thereby influence the verdict for the party they represented.

Mr Wilde opened the case for the prosecution by summarizing the indictment and pointing out that such a prosecution was brought as a last resort after accusations

---

15   Tal Golan, *Laws of Men and Laws of Nature* (Cambridge, MA, 2004), pp. 76–80.
16   *Manchester Guardian*, 24 August 1857.

that the works caused a nuisance had gone on over many years. He emphasized the important principle that for a successful prosecution it was necessary to prove that the nuisance had affected a large number of people and not just a few. He also accused Spence of constructing the tall chimney in June 1856 in an attempt to hide evidence of the nuisance. Both prosecution and defence were able to present their case with the aid of a model showing the layout of the alum works and the other manufactories surrounding it.[17] This model had been made for Peter Spence by James Fraser, an architect and surveyor of Pendleton. Evidence of damage to vegetation close to the alum works was provided by photographs commissioned from James Mudd by the landowner's solicitor to support the prosecution case, but newspaper reports have no record of their use in the proceedings.[18]

The prosecution then called a number of local residents to give evidence. As with so many other nuisance cases, the evidence was fragmented, contradictory and lacked consistency because it was based on each person's own experience, with some affected in their homes while others were not; some found their gardens affected to a greater or lesser extent; and some complained of a health ailment they attributed to the nuisance. Sir Frederick Thesiger, when cross-examining these witnesses, tried to concentrate on the recent period rather than on what had occurred several years earlier and to set their health ailments on a longer time scale.

The prosecution then called their main expert witness, Edward Frankland, to give evidence. Frankland was Professor of Chemistry at Owens College, Manchester and was well known to Spence and Angus Smith; indeed it was Frankland who had agreed, at Spence's request, to review the operation of the alum works some years earlier, and he had found no evidence of the alleged nuisance. It was also revealed that Frankland had made use of Spence's works over several years for the instruction of his students and there was mutual respect for each other's achievements. Frankland then summarized the chemistry of each process at the alum works, pointing out where the escape of nuisance gases was likely to occur. In the report commissioned by Spence and Dixon in November 1856, Frankland had found no evidence of any nuisance from the works. Subsequently he had undertaken his own independent investigation, but on this occasion not from inside the works but from outside, and had then attributed a number of nuisances including sulphur dioxide and hydrogen sulphide to the alum works. During his cross-examination, Sir Frederick Thesiger drew attention to the impropriety of not asking Spence's permission to visit the works, having been enlisted by the prosecution. Thesiger then read out in court a letter of 5 December 1856 in which Frankland had pointed out to Spence 'that while the works were conducted with the same care as they were in June, of the same year, he need not

---

[17]   This may be one of the first nuisance court cases in which a model was used.

[18]   The James Mudd collection of photographs is held at the Museum of Science and Industry in Manchester and the five photographs using in the Spence case are held at the Salford Local History Library.

fear any prosecution'.[19] Thesiger went on to add that Frankland 'did not think it necessary to inform the defendant he had subsequent to his June inspection found escapes of injurious gases; because he did not want to interfere with the impartiality and because he wished to maintain his incognito'.[20] Frankland's reply to this accusation of unprofessionalism is not recorded.

When opening the defence case, Thesiger started by pointing out the disadvantage of Spence not being able to give evidence. Even though his partner Thomas Dixon was giving evidence, Spence alone had intimate knowledge of the plant processes and their operation. Thesiger also reminded the jury that for the prosecution case to succeed it was necessary to prove that a large number of people had been affected by the alleged nuisance from the alum works, and highlighted the evidence of wealthy house owners who had bought properties close to the large manufactories but then complained of inconveniences.

The defence case was built around three areas: (1) effect of works on vegetation; (2) effect on health; (3) alleged discomfort. Thesiger then addressed Frankland's evidence as the report in the *Manchester Guardian* usefully summarizes:

> that report [Frankland's] was highly favourable to the defendant, and stated that there was no nuisance from the sulphuric acid, and that the means used for the destruction of the sulphuretted hydrogen was perfectly efficient. Now, however, Dr Frankland has become a deserter; and after partaking of the bounty of Mr Spence and taking advantage of his works for the instruction of his pupils, he had been induced to join the ranks of the enemy, and to get up evidence for the prosecution. Sir Frederick then described in sarcastic terms the mode in which Dr Frankland 'prowled' about the works; how ultimately he had succeeded in finding out some sulphurous acid and hydrocarbon; how instead of going to Mr Spence, to ask if any particular circumstances had occurred to account for the smell, Dr Frankland preserved what he called his incognito, in order that the inquiry might be 'impartially' conducted; and how at the same time that he was engaged for the prosecution, Dr Frankland wrote a bullying sort of note to the defendant, telling him that so long as he conducted his operations with the same care as he did in June, he need not fear any prosecution – which, however, was the very thing he had to fear secretly and clandestinely.[21]

Sir Frederick, having concluded his initial address, then called the defence witnesses. John Shaw, Secretary of the Manchester Botanical Society, had been asked by Mr Spence to inspect gardens within a radius of a mile of the works and was able to report that trees, hedges and gardens were healthy. Where he found damaged trees, this was due to fungi caused by bad drainage or due to the effects

19   *Manchester Guardian*, 24 August 1857.
20   Ibid.
21   Ibid.

of the wind (stag-headed), smoke or soot.[22] A number of local residents then gave evidence, highlighting the healthy state of their gardens, in contrast to witnesses for the prosecution. John Pickering, Inspector of Nuisances in Salford, and his sub-inspector, Benzamin Greenard, reported how they had inspected the works in 1855 and since then on a number of other occasions, and had never found that it created a nuisance.

Now it was the turn of Angus Smith, who started by referring to the report that he and Dr Crace-Calvert had prepared and how subsequently the reported defects were remedied. He then turned to the tests he had carried out to determine the amount of sulphurous acid passing up the chimney at Spence's works (53 grains per 1,000 cubic feet of air) compared with a sugar-refining chimney in Manchester (162 grains per 1,000 cubic feet of air). In an attempt to measure the acidity in the air close to Spence's works, Smith had collected rainwater samples from the neighbourhood of the alum works and compared them with 100 samples taken across different parts of Manchester: the former sample contained less than half the amount found in the latter.

At the conclusion of the defence case, and after Mr Wilde had reviewed the evidence presented by the defence, the judge started his summing up for the jury. He began by pointing out that the verdict should not be used to impede trade and commerce, and alum was very important to the manufacturing district in which the works was located, but 'on the other hand it was also important for gentlemen engaged in the production of such articles as alum, and other things, should not, for their own profit, annoy or injure their neighbours'. He went on to address the situation with Dr Frankland and commented 'that if the works had been carefully conducted, Dr Frankland became a witness for the defendant'.[23] The jury retired for an hour and three-quarters and then returned to hand the judge a note: 'Verdict, guilty as to the nuisance, which nuisance has not been proved to be prejudicial to health'.

This was a major blow to Spence, who had determinedly undertaken to ensure that his works did not cause a nuisance to the neighbours and had sought out pre-eminent professional advice. But this was not the end of the court proceedings, for on 23 November 1857 in the Court of the Queen's Bench Mr Wilde moved the judgement of the court. In reply Sir Frederick Thesiger pointed out that since the trial, Spence had further attempted to remove any grounds for complaint but the prosecution had not had sufficient time to examine this claim. The judges agreed to defer the judgement.[24] The case was considered again at a further meeting of the Queen's Bench on 3 May 1858, and it was agreed to let the judgement stand after Spence agreed to end the manufacture of alum at the Pendleton site on or

---

[22]   The smoke and soot were probably due to 'black smoke' from burning coal; this was also responsible for most of the sulphurous fumes in the air.

[23]   *Manchester Guardian*, 25 August 1857.

[24]   *Manchester Guardian*, 24 November 1857.

before 2 November.[25] After a 13-year fight Spence must have been worn out by the continuing uncertainty over the future of the Pendleton site. The works was moved to Miles Platting, where Spence must have hoped that any inadvertent nuisance would not raise the furore it had done at Pendleton.

Role of Expert Witnesses and Impact of 1857 Court Case on Angus Smith

The 1857 court case was important in a number of ways. Besides giving a clear insight into the working of the courts in such nuisance cases, it raised the issue of how the jury could weigh up the relative merits of the prosecution and defence witnesses, particular neighbouring residents likely to be most affected by the alleged nuisance. But there was the more challenging issue of how the judge and jury could assimilate the testimony from expert witnesses. As Mr Baron Channell said in his summing up:

> I think, when you come to sift and examine for yourselves the evidence of those whom I have called scientific witnesses, you will not find a great discrepancy between them. There are a great many points on which they agree, and the difference will really be with reference to those whom I have ventured to call the non-scientific witnesses.

Indeed, Mr Wilde in his summing up had spoken of largely desisting from cross-examining the scientific witnesses 'because he felt satisfied that scientific testimony was not that on which such a case ought to be determined; that must be of a character to speak directly to the facts'.[26]

These statements during the 1857 trial raised serious questions about the role of expert witnesses and how best to treat their evidence, issues that have reverberated to the present day.[27] Angus Smith was not only very disappointed with the outcome of the trial and the impact on his friend Peter Spence, but he was particularly shaken by how the expert evidence (scientific evidence in this case) was flagrantly manipulated by the inherent procedures of the judicial system. Smith had three primary objections to the current use of expert evidence. The first concerned the supposed conflict in the expert evidence by the opposing parties: there was often no conflict, but where it existed it was manipulated by counsels who had little or no understanding of the evidence and were prepared to distort it for their own selfish advantage. The second issue was at the heart of the evidence itself and the role of science in public affairs: if the evidence could be challenged at every

---

25   *Manchester Guardian*, 5 May 1858.

26   *The Daily Post*, 25 August 1857.

27   For recent discourses on the issue see Roger Smith and Brian Wynne, 'Introduction' and Brian Wynne, 'Establishing the rules of laws: constructing expert authority', in Roger Smith and Brian Wynne (eds), *Expert Evidence: Interpreting Science and the Law* (London, 1989), pp. 1–22 and pp. 23–55.

stage, then, Angus Smith argued, 'science would effectively be prevented from providing useful guidance in public affairs'.[28] The third issue was the overriding concern of the antipathy between science and the advocacy of law, comparing a scientist and a barrister. While a barrister in practice may be required to take a biased role, for the scientist it was different, as Angus Smith explained in his article 'Science in our Courts of Law':

> If the scientific man is the same [as the barrister] we have two parties with the same duties. A scientific man in that case simply becomes a barrister who knows science. But this is far removed from the idea of a man of science. He ought to be a student of nature, who loves whatever nature says, in a most disinterested manner. If we allow him or encourage him to be an advocate, we remove him from his sphere: we destroy the very ideal of his character; we give him duties which he never was intended to perform, and we turn him aside from the objects which first in early life led him to study in the direction of science.[29]

Angus Smith also felt strongly that expert evidence should be provided as written evidence and considered by the judge and barristers outside the adversarial court environment. Moreover, the ability of the judge and the jurors to assimilate detailed scientific evidence and weigh up its value in coming to their verdict remained another uncomfortable issue in trying to secure well-considered verdicts. Smith was infuriated by Frankland's behaviour – changing his position in the case and his treatment of Spence, a friend and close associate – and became increasingly concerned that expert evidence could be bought and, more worryingly, bought by the highest bidder. These were the issues Angus Smith pursued relentlessly over the next few years, taking opportunities to campaign through the National Association for the Promotion of Social Science, the Law Amendment Society and the Society of Arts.[30]

Angus Smith had previously given evidence to parliamentary inquiries and commissions but his appearance as a witness in the 1857 court case shook him severely, perhaps even traumatized him, causing him to re-examine many of his fundamental belief systems, as many of his subsequent writings and speeches show. Others were ready to support Angus Smith's stance: William Crookes wrote

---

28   Christopher Hamlin, 'Scientific Method and Expert Witnessing: Victorian Perspectives on a Modern Problem', *Social Studies of Science*, 16 (1986), p. 495.

29   R. Angus Smith, 'Science in Our Courts of Law', *Journal of the Society of Arts*, 7 (1860), pp. 136–7.

30   Ibid. Smith presented a paper at the first meeting of the Social Sciences Association in Birmingham in 1857 and at the second meeting in Bradford in 1858. The second paper was also read before the Law Amendment Society. In 1859 he raised the topic with the Chemistry Section of the British Association for the Advancement of Science at its meeting in Aberdeen. See also Golan, *Laws*, pp. 110–23.

in his *Chemical News*;[31] *The Times* and *Nature* devoted a considerable amount of space;[32] and from 1862 The British Association for the Advancement of Science had a long-standing committee chaired by Rev. Vernon Harcourt.[33] But others were ferociously opposed. One of the most vociferous was William Odling, Professor of Chemistry at Guy's Hospital (later at the Royal Institution, St Bartholomew's and Oxford), who felt that 'scientific witnesses are not infrequently opposed to one another on a question of fact, and very proper it is that they should be'.[34] With the differing modes of conduct of the law and science he felt that judicial procedures should not be changed and saw nothing wrong with the two interacting in a court of law. These debates were to continue into the twentieth century (and perhaps have never been resolved satisfactorily), but for Smith the trauma of the 1857 trial remained with him for the rest of his life and was to have a major influence on his work as Inspector of the Alkali Inspectorate.

Origin of 'The Monster Nuisance of All'

Many other nuisance cases occurred well before the Spence case in 1857, and many other industries besides alum were accused of making a nuisance, including copper works, but a serious offender from the 1820s, and the industry that was to galvanize action for parliamentary legislation, was the alkali industry, whose main product was soda (sodium carbonate). Until the 1770s supplies of soda were mainly from two natural sources – barilla, the ash of the plant *salsoda soda* cultivated along the Mediterranean coast of Spain centred on Alicante, and kelp, the *focus* seaweed found around the coast of the Western Isles of Scotland and of Ireland. Supplies of barilla and kelp were barely able to keep pace with the rapid rise in demand for soda due to expansion of the textile, soap and glass industries.

The French were particularly dependent on supplies of barilla, but these were often interrupted during disputes and wars with their continental neighbours. In 1776 the French Académie des Sciences offered a prize of 100,000 francs (worth about £4,000) for a good working method for making soda from common salt (sodium chloride). Although many of the eminent French chemists of the day

---

31   William Crookes, 'The Evidence of Experts', *The Chemical News*, 5 (1862), p. 281.

32   *Nature* devoted a number of articles to the topic; see 'The Whole Duty of a Chemist', 33 (1885), pp. 73–7; 'Scientific Assessors in Courts of Justice', 38 (1888), pp. 289–91.

33   For a report of the presentation by Rev. Vernon Harcourt to Section F – Economic Science and Statistics, see 'The British Association for the Advancement of Science', *Chemical News* (11 October 1862), pp. 189–90, and 'Report of the Committee on Scientific Evidence in Courts of Law', *Report of the Nottingham Meeting of the British Association for the Advancement of Science in 1866* (London, 1867), pp. 456–7.

34   William Odling, 'Science in Courts of Law', *Journal of the Society of Arts*, 8 (1860), p.168.

tried to find a solution, it was Nicholas Leblanc, surgeon to the Duc d'Orléans, who submitted the most satisfactory proposal. A schematic representation of the process is shown in Figure 5.1. Unfortunately, with the tumult and uncertainty caused by the French Revolution Leblanc never received his prize, although his process was granted a patent in 1791, the same year that he began manufacturing soda at his St-Denis works.[35]

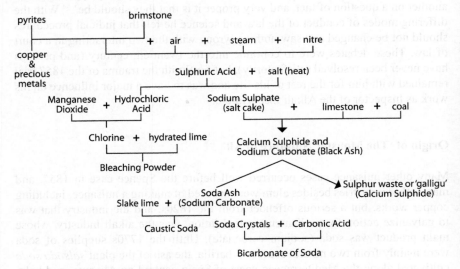

Figure 5.1   Diagram of the Leblanc process

In the aftermath of the French Revolution many British people travelled to France, and in particular Paris, to see how the momentous events had changed French life and society. William Losh, a chemical manufacturer on Tyneside who had studied chemistry at Cambridge, spent time in Paris working with the French chemist Antoine Lavoisier. He gathered information on some of the new-thinking chemical understanding and manufacture, including the Leblanc process and visited Leblanc's works at St-Denis during the Peace of Amiens (27 March 1802 and 17 May 1803). Returning to Tyneside, Losh initially continued with the old method of producing soda – using salt and lead oxide (litharge) – but started experimenting with the new process. After further visits to St-Denis between 1802 and 1806 he started working with the Leblanc process about 1814.

---

35   For an account of Leblanc's role in developing the process, see Ralph E. Oesper, 'Nicholas Leblanc (1742–1806)', *Journal of Chemical Education*, 19 (1942), pp. 567–72. Leblanc never received the prize money and committed suicide in 1806. Although Napoleon III was petitioned by Leblanc's heirs in 1855, no payment was forthcoming. See Ralph E. Oesper, 'Nicholas Leblanc (1742–1806)', *Journal of Chemical Education*, 20 (1943), pp. 17–19.

Figure 5.2   James Muspratt (1793–1886), chemical manufacturer and Father of
               the British Heavy Chemical Industry
               J. Fenwick Allen, *Some Founders of the Chemical Industry* (London,
               1906), facing page 69

Other manufacturers in Britain followed Losh's lead. From the late eighteenth
century Scotland (and in particular Clydeside) had become an important centre for
the alkali trade. Charles Tennant was a weaver and bleacher before building his
St Rollox works on the Monkland Canal near Glasgow, where he was producing
about 100 tons of Leblanc soda by 1818. In Liverpool Thomas Lutwyche and
William Hill may have operated the Leblanc process from 1814, although there

remains some doubt about this enterprise.[36] A major step up in the level of production occurred from 1823 when James Muspratt, having arrived in Liverpool from Dublin the previous year, started using the Leblanc process at his Vauxhall Road works alongside the Leeds–Liverpool Canal in Liverpool. Later Muspratt became known as the 'Father of the British Heavy Chemical Industry' because of his leadership in manufacturing large-tonnage chemicals such as soda and sulphuric acid.[37] Merseyside, Tyneside and Glasgow became the key centres for the Leblanc soda trade, with their ready access to large quantities of salt and coal and easy transport for their chemical products. Merseyside had ready supplies of salt (from Cheshire), coal (from St Helens) and limestone (from North Wales and Derbyshire), all easily transported to the production sites and with access to the major Atlantic port of Liverpool. It was this triangular trade and the economic advantage that the Leblanc soda could provide for the soapboilers that persuaded Muspratt to take the opportunity to move to Liverpool in 1822.[38]

While the Leblanc process provided the means to produce large quantities of soda, it was plagued by harmful chemical by-products that manufacturers struggled to reuse. Hydrogen chloride gas (or muriatic acid gas, as it was known in the alkali trade) was produced during the first stage of the process when sulphuric acid was reacted with salt. The other harmful by-product was sulphur waste (or 'galligu', the black, viscous, smelly residue), produced during the third stage when the black ash (resulting from the heating of saltcake (sodium sulphate), coal and limestone) was 'lixiviated', or vigorously mixed, with water. In the early years of the process in Britain there was no use for either of these waste products and so the hydrogen chloride gas was released into the atmosphere from tall chimneys and the galligu was deposited on waste ground surrounding the works. Both caused injury to the natural and human environments, adding to the debilitating conditions already created by the black smoke from burning large quantities of coal.

All Leblanc manufacturers faced nuisance accusations arising from the hydrogen chloride gas almost from the start of their operations; the frequency and

---

[36] For details of the Tennant family and the early history of the St Rollox works, see L.F. Haber, *The Chemical Industry during the Nineteenth Century* (Oxford, 1958), pp. 14–17, and E.W.D. Tennant, 'Early History of the St. Rollox Chemical Works', *Chemistry and Industry*, 66 (1947), pp. 667–73. For an account of the problem of acid gas at the St Rollox works, see Peter Reed, 'Where Even the Birds Cough: The First British Cases of Large-scale Atmospheric Pollution by the Chemical Industry on Merseyside and Clydeside in the Early 19th Century', in M. Fetizon and W.J. Thomas (eds), *The Role of Oxygen in Improving Chemical Processes* (6th BOC Priestley Conference) (Cambridge, 1993), pp. 115–22.

[37] D.W.F. Hardie and J. Davidson Pratt, *A History of the Modern British Chemical Industry* (Oxford, 1966), p. 26.

[38] For a discussion of the triangular trade, see T.C. Barker, 'Lancashire Coal, Cheshire Salt and the Rise of Liverpool', *Transactions of the Historic Society of Lancashire and Cheshire* (1951), pp. 83–101.

vociferousness of the accusations varied depending on the location of the works and the nature of the surrounding area. Losh faced a petition from Newcastle Town Council that graphically describes the damaged from acid gas:

> the gas from these manufactories is of such a deleterious nature as to blight everything within its influence, and is alike baneful to health and property. The herbage of the fields in their vicinity is scorched, the gardens neither yield fruit nor vegetables; many flourishing trees have lately become rotten naked sticks … It tarnishes the furniture in our houses, and when we are exposed to it, which is of frequent occurrence, we are afflicted with coughs and pains in the head … All of which we attribute to the Alkali works.[39]

But accusations were to continue against all the alkali works on Tyneside using the Leblanc process, which by the 1820s was all of them.

The position with Tennant in Glasgow and Muspratt in Liverpool provides an interesting contrast in how the manufacturers were confronted by nuisance accusations and how they dealt with these public protests. Tennant and his St Rollox works managed to avoid the worst of the complaints because the works was to the north of Glasgow, and the prevailing south-west winds carried the black smoke and hydrogen chloride gas away from populated areas. Like Muspratt, Tennant relied on a very tall chimney to carry the smoke and gas away; the chimney, affectionately known as 'Tennant's Stalk', was just over 400 feet high and was reputed to be the tallest in Britain. Its construction was a remarkable feat of civil engineering:

> The chimney is founded upon a bed of solid sandstone rock, twenty feet below the surface of the ground. The diameter of the outer chimney is fifty feet at its foundation, forty feet diameter at the surface of the ground, and will diminish … to a diameter of fourteen feet six inches, when it will have attained an altitude of from four hundred and twenty to four hundred and thirty feet. The inner chimney is a cylinder of sixteen feet diameter, rising perpendicularly to a height of two hundred and sixty feet. This inner chimney is unconnected with the outer one, but comes very nearly in contact at its termination, allowing only space for expansion arising from temperature. The flues from the various parts of the extensive works are introduced into the inner chimney through four circular apertures, each seven feet six inches in diameter.[40]

---

[39]   *Proceedings of the Town Council of Newcastle upon Tyne*, 9 January 1839, p. 19, Tyne and Wear Archives.

[40]   *The Mirror*, 4 December 1841.

Figure 5.3   James Muspratt's Vauxhall Road works, Liverpool, c. 1830, print in
                  the Brierley collection
                  Reproduced by Permission of Liverpool City Libraries

For Muspratt, the situation turned out very differently, even though he too used a tall chimney to disperse the hydrogen chloride gas. The chimney at the Vauxhall Road works was 225 feet high, which made it a very prominent landmark (see Figure 5.3). Even though there were others in the surrounding area, none stood out so prominently. Also, the Muspratt works was a recent addition to the local industrial scene that seemed to coincide with the alleged nuisance. More significantly for the nuisance accusations, the district of Everton was about a mile from Vauxhall Road in a north-westerly direction and at about 225 feet above the alkali works, and therefore at about the same height as the top of the chimney. With the steady increase in production and the amount of hydrogen chloride gas released, so the accusations mounted.

A letter headed 'Alleged Nuisance' in The *Liverpool Mercury* of 5 October 1827 is typical:

> GENTLEMEN – The chymical works of Mr. J. Muspratt, 107, Vauxhall road, pour forth such volumes of sulphurous smoke as to darken the whole atmosphere in that neighbourhood, so much so, that the Church of St. Martin-in-the-Field, now erected, cannot seen from the houses at about one hundred yards distance, the stones of which are already turning a dark colour from the same cause. The scent is almost insufferable, as well as injurious to the health of the persons residing in that neighbourhood. I should think this parish is so much interested in the afore-mentioned church, which will shortly be changed from white stone to black, as to prevent the continuance of so great a nuisance.

A CONSTANT READER.[41]

As with the Spence case of 1857, it was not long before court proceedings started.

Difficulty Getting Successful Prosecutions

By examining some of the early Leblanc nuisance court cases it is possible to appreciate the level of alleged damage caused, and more especially some of the difficulties with obtaining a successful prosecution within the existing legal framework. The first case indicting Muspratt and his Vauxhall Road works was in April 1828, when a group of churchwardens in Liverpool brought an action for 'creating a nuisance'.[42] This was not the first time a manufacturer in Liverpool had faced such action, for in 1770 Charles Roe, a copper smelter, lost his court case and his copper works was 'excluded from Liverpool'.[43] Muspratt defended his

---

[41] 'Alleged Nuisance', *Liverpool Mercury*, 5 October 1827.

[42] *Liverpool Mercury*, 1 August 1828.

[43] Gordon W. Roderick and Michael D. Stephens, 'Profits and Pollution: Some Problems facing the Chemical Industry in Liverpool in the Nineteenth Century. The

operations and the means of dispersing the acid gas, and although he lost the case, he was fined only one shilling, a financial penalty unlikely to cause him much reflection or concern. Muspratt's response was to raise the height of his chimney in the hope of sending the acid gas well above Liverpool and the surrounding area.

This case proved to be just the opening salvo as accusations continued. In April 1831 the inhabitants of Everton brought an indictment against Muspratt for 'nuisance arising from his works'. Many prosecution witnesses gave evidence about the injurious nature of the acid gas and its unpleasant smell. Initially these witnesses were from Liverpool, but when witnesses from Everton were called, the defence challenged their eligibility since the indictment was laid in Liverpool, not in Everton, and the judge acceded. Undaunted, the prosecution proceeded to introduce medical and scientific evidence on the injurious nature of the acid gas and its effects on vegetation and on the health of animals. The defence countered with witnesses who had not experienced these effects. Several local soap manufacturers (possibly customers of Muspratt) gave evidence on the economic advantages of the Leblanc soda, as did doctors (including the Muspratts' family doctor), recommending 'the erection of such works in every town for the purpose of purifying the air'.[44] The jury returned with a verdict for the defence.

The verdict prompted an advertisement in the *Liverpool Mercury* the following week, promoting a Muriatic Acid Gas Joint Stock Company so that local inhabitants could benefit:

GRAND DISCOVERY

NEW JOINT STOCK COMPANY

As it has been satisfactorily proved on the late trial at Kirkdale – The King versus Muspratt, that

MURIATIC ACID GAS

so far from being deleterious, as has been erroneously imagined, is highly beneficial to health and longevity – it is proposed to establish A MURIATIC ACID GAS JOINT STOCK COMPANY, to supply the town, by means of pipes, with this most valuable article. Immediate application would be politic, as the shares will, most assuredly, soon be at a high premium.[45]

Banners supporting the company appeared on chimneys.

Corporation of Liverpool versus James Muspratt, Alkali Manufacturer, 1838', *Industrial Archaeology*, 11/2 (1974), p. 36.

[44]   'Report of the Proceedings', *Liverpool Journal*, 7 May 1831.

[45]   'Announcement', *Liverpool Mercury*, 13 May 1831.

The case brought into the open for the first time some of the important issues that future court cases would focus on even more sharply: the scale of damage attributable to acid gas; attributing responsibility for damage to one particular works; the use of scientific and medical evidence; the economic benefits of the soda industry to both the national and the Liverpool economies.

By 1831 Muspratt had expanded his business with additional works at St Helens and Newton-le-Willows. The St Helens works was at Gerard's Bridge and was developed through a partnership between Muspratt and a friend from Dublin, Josias Gamble, but with nuisance accusations mounting the partnership was soon terminated. Gamble remained in St Helens while Muspratt started to develop a major new works at Newton-le-Willows at a site in a rural setting alongside the Sankey Canal (later St Helens Canal).

In Liverpool, Muspratt was still being harassed and in April 1838 he was indicted by the Corporation of Liverpool for 'creating and maintaining a nuisance within the Borough, to the annoyance and injury of the inhabitants thereof'. The case was heard at the Liverpool Spring Assizes before Sir John Taylor Coleridge (nephew of the poet Samuel Taylor Coleridge). Extensive reports of the proceedings appeared in *The Times* as well as in local newspapers. It became something of a *cause célèbre* and 'excited the greatest interest in Liverpool and the vicinity'.[46] A full transcript of the proceedings was published soon after the trial and provides probably the best account of such legal proceedings.[47] Although Roderick and Stephens have claimed that this case was 'the most decisive and the most significant', it was just part of the continuum of cases that extended well into the 1850s.[48]

The prosecution called 49 witnesses, including four scientific and medical witnesses, while the defence called 46 witnesses, including the manager and foreman of the Vauxhall Road works. Thomas Thomson, Professor of Chemistry at the University of Glasgow and a friend of Muspratt, appeared as a scientific witness for the defence. The proceedings followed a similar pattern as in earlier cases but the scientific and medical evidence was examined in greater detail and the economic arguments were put more forcefully by the defence. The scientific and medical evidence provided by each side's expert witnesses and the direction of cross-examination proved difficult for the judge and jury to understand. When summing up the evidence for the jury at the conclusion of the case, the judge anticipated their difficulties:

---

[46]   'Report of Queen v Muspratt', *The Times*, 10 April 1838.

[47]   *A Full Report of the Trial of the Important Indictment Preferred by the Corporation of Liverpool Against James Muspratt, Esq., Manufacturer of Alkali at the Liverpool Spring Assizes 1838 before Sir John Taylor Coleridge, Knight, and Special Jury, for a Nuisance Alleged to Proceed from his Chemical Works in Vauxhall Road, Liverpool* (Liverpool, 1838). Copy in Special Collections, Harold Cohen Library, University of Liverpool.

[48]   Roderick and Stephens, 'Profits and Pollution', p. 35.

I shall show you the effect of this chemical testimony more than anything else, as I am entirely ignorant of chemistry, and if it should happen that any of you have any knowledge of chemistry, I would not advise you to place any reliance upon it, as a smattering in chemistry is exceedingly dangerous.[49]

A key part of the defence strategy was the economic benefits brought by the Leblanc soda industry, a position Muspratt would have argued for strongly as a Liberal and a staunch free trader. Although the Muspratt legal team made a determined effort to enter such evidence, the judge considered this inadmissible and in his concluding remarks elaborated his reasoning:

With regard to the consequence of this trial upon Mr. Muspratt, or upon the trade of the country, if you could for a moment imagine that you would by your verdict, banish from this country to foreign shores a very thriving branch of the trade, all this has nothing to do with the subject unless to make you weigh your verdict with the greatest care. If Mr. Muspratt is guilty of a nuisance, it is no defence in law that he confers great benefit on other persons, and still less could it be argued that in so doing he has embarked a large capital, or has accumulated to himself great and honourable fortune. The real question at issue is whether Mr. Muspratt is guilty of what is charged against him.[50]

The record of the trial also highlights the prevailing difficulty of attributing responsibility for the damage to one works, for the Muspratt works was not in isolation but surrounded by other manufactories. The Vauxhall Road area was more like an 'industrial estate', which was made clear when Samuel Home produced a detailed plan of the area around Vauxhall Road showing the location of other alkali works and manufactories numbering over 100; these included:

12 chemical works, 23 distilleries, 17 soaperies, 16 brewers, 7 lime works, 17 foundries, 2 gas works, tan yards, 3 sugar houses, 3 colour manufactories, 5 water works, 13 steam mills, coachmakers, mortar mills, etc.[51]

What rich effluvia must have emanated from this area, and no wonder it was often referred to as the 'Island of Spices'.[52] Much of the evidence attributing responsibility to Muspratt's works came from witnesses in Everton who claimed to have 'followed with their noses, the vapour, which they traced to Mr. Muspratt's works', a rather dubious method of verification.[53] What was needed was a more rigorous scientific approach that eliminated speculation and uncertainty.

---

[49]   *A Full Report of the Trial*, p. 95.
[50]   Ibid.
[51]   Ibid., p. 13.
[52]   Ibid., p. 41.
[53]   Ibid., p. 43.

On this occasion the jury returned a verdict against Muspratt; this had little impact on the operation of the Vauxhall Road works, although Muspratt may have increased the height of the chimney again. By this time he was concentrating his efforts much more on Newton-le-Willows, where soda production increased rapidly to meet the demands of the new markets in North America.

Intervention of Landed Gentry and Landowners

While the works in Liverpool was in the centre of town and surrounded by people, the site at Newton-le-Willows was in prime countryside. In such a rural setting Muspratt had probably hoped to put behind him all the nuisance accusations of recent years. However, by 1836 and before the increase in soda production for the North American market, concerns were being expressed by the local community.[54]

The Rowson and Cross Papers at the Lancashire Record Office list 19 separate claims for damages against Muspratt and his Newton works between 1846 and 1862.[55] The claims vary in scale: small tenant farmers claimed damages of between £50 and £100, while the major landowners were claiming damages of between £20,000 and £30,000. The total number of claims was well in excess of 19, since many claims for small amounts were settled promptly by Muspratt's agent directly with the claimants to avoid antagonizing the local community and the cost of lengthy court litigation. With small claims being settled without detailed evidence, suspicions mounted that some were fraudulent. Nevertheless, the larger claims for compensation were defended vigorously.

One such claim was brought in 1846 by Sir John Gerard of New Hall, near Garswood, Lancashire. The Gerard family had owned large tracts of land in Lancashire since 1611, and the claim referred to damage to woodland plantations.[56] The damage had become noticeable in 1837, with a further marked deterioration in 1839–40. Expert opinion linked the damage to acid gas from Muspratt's works at Newton-le-Willows. Muspratt's lawyers tried to mitigate the alleged damage and thereby reduce the level of compensation that might have to be paid. Three lines of approach were taken: the effective dispersal of the acid gas (supported by detailed scientific evidence on the concentration of acid gas entering the chimney); poor management of the woodlands; and the varying effects of acid gas on different

---

54   Letter to James Muspratt from James Hornby (Winwick), dated 4 November 1836. LRO (Muspratt Papers, 920MUS/2-39).

55   Rowson and Cross Papers at the Lancashire Record Office (Ref: DDC/18) contain a series of files on court cases brought against Muspratt and his Newton works. Included are the court cases brought by Sir John Gerard. Rowson and Cross were solicitors in Prescot.

56   From the sixteenth century the Gerard family were extensive landowners. The first baronet was created on 22 May 1611; Sir John Gerard was the 12th Baronet of Bryn. In 1876 Sir John's brother, Robert, was created 1st Baron Gerard of Bryn. See Burke's *Peerage and Baronetage*, pp. 1131–2.

species of trees. The jury returned a verdict for the plaintiff but Muspratt was probably very relieved that the damages were reduced from £30,000 to £1,000.[57] The Gerard court case highlights a significant shift: Leblanc soda manufacturers such as Muspratt were starting to face the wrath of more powerful and influential opponents than before – the country landowners and landed gentry – and the intensity of the opposition would only grow.[58]

Select Committee on Injury from Noxious Vapours 1862 and Alkali Act 1863

By 1862 the Leblanc soda industry had become a major contributor to the national economy: consuming 1.8 million tons of raw materials, employing 19,000 workers, producing 3.8 million tons of waste products, and involving £2 million working capital.[59] But the industry was also responsible for releasing vast quantities of hydrogen chloride gas from their chimneys and this proved almost impossible to regulate. As more and more wealthy landowners suffered the effects of the gas and found their property values diminished as a consequence, so the pressure for parliamentary action grew. Lord Derby with his large estate at Knowsley Hall (7 miles to the east of Liverpool) had personal experience of the damage and when a deputation of landowners and members of the House of Lords made representations to him in 1862 he decided to take urgent action. As a former prime minister, Derby knew the political system intimately and soon after receiving the deputation persuaded the House of Lords to set up the Select Committee on Injury from Noxious Vapours. *Punch* humorously referred to it as 'Lord Derby's Smell Committee'.[60] Events moved quickly: on 9 May the motion was heard; on 12 May the Committee was named with Derby as chairman; and on 16 May the first witnesses were called.[61] Committee members were carefully selected: eminent scientists such as Lyon Playfair, Edward Frankland and Wilhelm Hofmann gave evidence alongside land agents, doctors and manufacturers. From the outset Derby

---

[57]   'Report of Gerard, Bart. v Muspratt', *The Times*, 3 September 1846.

[58]   For a discussion of the role of the landowners in the problem of pollution from alkali works, see A.E. Dingle, '"The Monster Nuisance of All": Landowners, Alkali Manufacturers, and Air Pollution, 1828–64', *Economic History Review*, 35 (1982), pp. 529–47. For an account of the growing opposition to this pollution, see Sarah Wilmot, 'Pollution and Public Concern: The Response of the Chemical Industry in Britain to Emerging Environmental Issues, 1860–1901', in Ernst Homburg, Anthony S. Travis and Harm G. Schröter (eds), *The Chemical Industry in Europe, 1850–1914: Industrial Growth, Pollution and Professionalization* (Dordrecht, Netherlands, 1998), pp. 121–47.

[59]   *First Report of the Inspector appointed under the Alkali Act*, P.P. 1865 (3460), pp. 159–60.

[60]   *Punch*, 24 May 1862, p. 204.

[61]   Roy M. MacLeod, 'The Alkali Acts Administration, 1863–84: The Emergence of the Civic Scientist', *Victorian Studies*, 9 (1965-66), p. 88.

appeared to have a plan (albeit probably in his mind initially) as to the desired outcome and the nature of the parliamentary legislation to follow.[62]

One of the witnesses called to give evidence to the Select Committee was William Gossage, an alkali manufacturer with 30 years' experience. Gossage had gained a good grounding in practical chemistry as well as an understanding of the theoretical aspects of the subject through reading several of the standard texts of the day while apprenticed to his uncle, a druggist in Chesterfield. By 1831 Gossage had moved to Stoke Prior in Worcestershire and become a large-scale manufacturer of chemicals and soap as the British Alkali Company, making a range of chemicals using local sources of brine.[63] Like Muspratt, Gossage had received complaints from local people over the 'nuisance' from acid gas since he too was using the Leblanc process.

From a young age Gossage had shown an aptitude for creative invention and, when faced with the problem of the hydrogen chloride gas, he realized that he had to find a working solution if his business was to continue to flourish. Adjacent to the Stoke Prior works he found a derelict windmill, packed it with gorse, brushwood and coke, and then allowed a stream of water together with the acid gas to enter at the top.[64] Since acid gas is very soluble in water, nearly all the acid gas was absorbed by the water by the time the latter had reached the bottom of the mill.[65] Details of his invention of the acid tower, as it became known, were included in Gossage's patent 'Apparatus for Decomposing Salt, etc' (B.P. 7267/36).[66] The patent describes several of his subsequent modifications that made the acid tower even more effective. By means of the acid tower Gossage was able to use his tall chimney for its original purpose – to provide a draught between the saltcake pan and the atmosphere.

---

[62]   Ibid. The 14 members of the House of Lords comprising the Select Committee included two scientists, five estate owners (including Derby), the Comptroller of the Treasury, and two others from Derby's previous government administration.

[63]   See plans (with a schedule) of the British Alkali Company at Stoke Prior works. WCRO (Ref: BA 8851/20). For information about the British Alkali Company, see Alan White, *Worcestershire Salt: A History of the Stoke Prior Salt Works* (Bromsgrove, 1996), pp. 13–20 and also Alan White, *The Worcestershire and Birmingham Canal: Chronicles of the Cut* (Studley, Warwickshire, 2005), pp. 265–8.

[64]   For the type of windmill Gossage used, see Peter Reed, 'Acid Towers and the Control of Chemical Pollution 1823–1876', *Transactions of the Newcomen Society*, 78 (2008), p. 108.

[65]   One litre of water dissolves 500 litres of hydrogen chloride, yielding a solution containing 42 per cent of hydrogen chloride at one atmosphere pressure and 0 °C (standard conditions).

[66]   B.P. 7267/36 concerned 'Certain improved apparatus for decomposing common salt, and for condensing and making use of the gaseous product of such decomposition, also certain improvements in the mode of conducting these processes'. The condensing tower is described in detail with a summary of its operation; some drawings are included.

While giving evidence, Gossage was asked about the general adoption of the acid tower within the trade, to which he was able to reply:

> Almost all the trade are using it now, some more carefully than others; those who used it carefully do not do any mischief; those who use it carelessly of course do mischief.[67]

But the acid tower was not widely adopted within the industry between its invention in 1836 and the late 1850s – a period during which acid gas continued to create a nuisance and court cases were pursued with vigour. Why did this situation occur?

There were several reasons why the acid tower was not adopted generally until the late 1850s: the underlying scientific principles of gas absorption by liquids were not understood; there was no ready use for the hydrochloric acid (from the acid tower); the cost of constructing acid towers was perceived as prohibitive; and there was no legal enforcement to use them. The science of gas absorption in liquids was not understood. While the effect of temperature on the solubility of gases in liquids was well known following the work of William Henry in the early 1800s, little was known about the physical interface between a liquid and a gas to maximize absorption.[68] It appears that most chemical manufacturers understood that 'bulk of liquid' was the determining factor in absorbing large quantities of gas. It was this prevailing view that led Muspratt to make his famous outburst on hearing details of Gossage's invention: 'Sure, all the waters of Ballyshannon itself would not suffice to condense the acid I make.'[69] So why did Gossage's windmill work so effectively?

Gossage came to understand that the bulk of water was not the key factor in dissolving the gas but rather the surface area of contact between the acid gas and the water. As the water percolated down the mill, it passed over the gorse, brushwood and coke, producing a thin film of large surface area of contact between the acid gas and the water.[70] Other key factors in maximizing condensation of gases are contact time between the water film and the gas, and the temperature of the gas.

---

[67]   *Select Committee on Injury from Noxious Vapours*, p. 85.

[68]   William Henry, 'Experiments on the Quantity of Gases absorbed by Water, at different Temperatures, and under different Pressures', *Philosophical Transactions*, 93 (1803), pp. 29–42.

[69]   J. Fenwick Allen, *Some Founders of the Chemical Industry* (London, 1906), p. 88.

[70]   Scrubbers were used in coal-gas manufacture to remove ammonia and ammonium salts from the gas, but their use does not pre-date Gossage's invention. There is no mention of scrubbers in the 1841 edition of Samuel Clegg, *A Practical Treatise on Manufacture and Distribution of Coal Gas* (London, 1841), but the 1853 edition does have a section on scrubbers; see pp. 198–9. Today the principle of the acid tower is used in scrubbers or scrubbing towers in the petrochemical industry; see McGraw-Hill *Encyclopaedia of Science and Technology*, 6th edn (vol. 1, 1987), p. 15.

Another reason why the acid tower was not adopted more quickly was that there was no use for the large quantities of hydrochloric acid produced. During his evidence to the 1862 Select Committee, Lyon Playfair, a frequent adviser to government on scientific matters, agreed that only 20 per cent of the hydrochloric acid could be sold (to produce chlorine for bleaching powder). The remaining 80 per cent was run as waste into streams, canals and rivers adjoining the alkali works – one form of pollution converted into another form! Besides destroying life in these waterways, the acid, if it came in contact with sulphur waste (the other principal waste product of the Leblanc process), hydrogen sulphide, a toxic gas with its characteristic smell of bad eggs, was produced, adding to the debilitating environmental conditions in the vicinity of the works. Resolution of the hydrochloric acid dilemma came only in the early 1860s following the sharp increase in demand for bleaching powder when esparto grass was imported to supplement rags in papermaking.[71]

The cost of constructing acid towers and the lack of enforcement to use them are closely linked. Manufacturers were reluctant to commit heavy expenditure on plant when it could be avoided. The additional cost of building and operating new plant might make the manufacturer's products less competitive compared with those of manufacturers who did not have acid towers. There was no legal enforcement that would place equal responsibility on all alkali manufacturers.

It was left to each manufacturer's individual circumstances (including the impact of court fines and claims for compensation) to determine who adopted acid towers as an integral part of the operation of the Leblanc process and who did not. The Select Committee was determined to remove this decision-making choice from manufacturers.

During the Select Committee hearing several witnesses, including William Gossage and Lyon Playfair, were questioned about the need for regular inspection of the alkali works. This would necessitate physical access to the working plant in the alkali works and require the individual manufacturer's permission. Gossage expressed support for control legislation:

> The feeling among those who conduct their business properly is they would be exceedingly glad to have legislation to compel gas to be condensed, because then those at present doing their work in a slovenly way would conduct their business properly, and those who manage their works a proper way would be relieved from the consequences of the other misdeeds.[72]

However, many manufacturers were against any form of government interference in the workings of industry and any change in the *laissez-faire* approach of the

---

[71]   British imports of esparto grass rose from 16 tons in 1861 to 200,000 tons in 1887; see Haber, *The Chemical Industry*, p. 96.

[72]   *Select Committee on Injury from Noxious Vapours*, p. 85.

time. Any change for the alkali industry would set an unfortunate precedent for the future relationship between government and industry. As the Select Committee proceedings continued, it must have become increasingly clear to the manufacturers that a point had been reached at which the Select Committee report would proceed in due course to parliamentary legislation. In these circumstances the manufacturers agreed to show support for the work of Lord Derby and his Committee, but also to try to influence its recommendations. By this approach it might be possible for the manufacturers to ameliorate any constraints on the day-to-day work of the alkali industry and to rescue a reputation that had been blighted over the previous 40 or so years.

During the Select Committee proceedings John Hutchinson, an influential alkali manufacturer in Widnes, a town that later became known as 'the chemical town', used his influence with other alkali manufacturers. During the 1850s Hutchinson had purchased large tracts of land in Widnes and the land was subsequently leased to other chemical entrepreneurs such as Muspratt, Gossage and Deacon.[73] Following his consultation, Hutchinson submitted a statement on behalf of the majority of his fellow manufacturers, part of which stated:

> The majority of the trade is willing to concur in the object proposed by Lord Derby in his speech, namely, the compulsory condensation of Muriatic Acid Gas, provided such time is given for the consideration of the subject, as will enable a measure to be framed which, while protecting the public, will not be injurious to a manufacture occupying such so large an amount of capital and labour, so important to the prosperity of the country at large, and essential to the actual existence of large communities ... If this course be adopted, the majority of the trade will give its best assistance and co-operation in the preparation and carrying out of such a measure.[74]

The initial pace set by Lord Derby was maintained through to approval of the legislation. *The Report of the Select Committee* was published in August 1862, and a private bill was introduced in March 1863 by Lord Stanley of Alderley (a relative of Lord Derby).[75] The bill was passed and received its Royal Assent in July 1863, demonstrating that when Parliament needs to work quickly, it can do so with the right leadership, direction and commitment. The terms of the legislation were effective from 1 January 1864.

---

[73] It is interesting to note that while Hutchinson and the other alkali manufacturers in Widnes used the Leblanc process, it was two of his employees, John Brunner as office manager and Ludwig Mond as a technical adviser, that later formed the successful partnership of Brunner Mond and Company Ltd and adopted the ammonia-soda process as a direct competitor to the Leblanc process.

[74] *Select Committee on Injury from Noxious Vapours*, p. x.

[75] Lord Stanley of Alderley and Lord Derby were related through the Stanley family name. See entries in Burke's *Peerage and Baronetage*.

The legislation was passed as the Alkali Act (1863) for an initial period of five years and established the Alkali Inspectorate within the Board of Trade. Its terms included: registration of all alkali works; the requirement for at least 95 per cent of the acid gas to be condensed; the appointment of an inspector who reports annually to Parliament; and the recovery of damages by civil action brought by the inspector in the County Court. However, for many manufacturers these terms represented a serious intrusion: 'The terms of the legislation reflected the pleas of the manufacturers for moderation. They saw the government as interlopers, with a question mark over their impartiality.[76]

It was the inspector's responsibility to implement the terms of the legislation, but a good working relationship with the manufacturers was essential to achieve consistent enforcement.

---

[76]   Peter Reed, 'Robert Angus Smith and the Alkali Inspectorate', in E. Homburg, A.S. Travis and H.G. Schröter (eds), *The Chemical Industry in Europe, 1850–1914: Industrial Growth, Pollution, and Professionalization* (Dordrecht, Netherlands, 1998), p. 157.

The legislation was passed as the Alkali Act (1863) for an initial period of five years, and established the Alkali Inspectorate within the Board of Trade. Its terms included; registration of all alkali works; the requirement for at least 95 per cent of the acid gas to be condensed; the appointment of an inspector who reports annually to Parliament; and the recovery of damages by civil action brought by the inspector in the County Court. However, for many manufacturers these terms represented a serious intrusion. The terms of the legislation reflected the plan of the manufacturers for moderation. They saw the government as interlopers, with a question mark over their impartiality.

It was the inspector's responsibility to implement the terms of the legislation, but a good working relationship with the manufacturers was essential to achieve consistent enforcement.

Peter Reed, *Robert Angus Smith and the Alkali Inspectorate*, in E. Homburg, A.S. Travis and H.G. Schröter (eds), *The Chemical Industry in Europe, 1850–1914: Industrial Growth, Pollution and Professionalization* (Dordrecht: Reidel, 1998), p. 157.

Chapter 6
The 'Civil Scientist' in Action

Between the time the Alkali Act (1863) received parliamentary approval in July, and came into effect as from 1 January 1864, the Board of Trade, as the responsible government department, created the Alkali Inspectorate. The legislation had no precedents as far as operating procedures were concerned; although other inspectors were working for government departments, such as the factory inspectors, mining inspectors and railway inspectors, the nature of the alkali inspectors' work was very different and their powers were very much greater. Robert Angus Smith, as the first Inspector of the Alkali Inspectorate. had a blank slate upon which to devise the operating procedures.

Besides the detailed monitoring procedures for the works, Angus Smith was aware of the concerns shared by the different parties in the outcome of the Alkali Inspectorate's work. There was Parliament that had set very clear objectives; the government that had taken on the responsibility through the Board of Trade and could ill afford to be seen to take a lax approach to controlling the debilitating pollution; the manufacturers who during the Select Committee had remained split over giving the inspectors unbridled access to their works; and the public, who had protested for many decades and were now sceptical about the ability to effectively control the pernicious acid gas. Nevertheless, with Angus Smith's appointment in February 1864, because of his standing as the pre-eminent expert on the factors controlling air quality, there were high expectations of the inspection team. Only time and experience would tell whether Angus Smith could rise to the challenge not only of ensuring that less than 5 per cent of the acid gas was released from the works but also, perhaps, balancing the expectations of the various parties, an even more demanding challenge.

Blank Slate for Regulation

Before the Act came into effect on 1 January 1864, the Board of Trade and in particular Angus Smith had to draw up the procedures to support the terms of the legislation. The first priority was the composition of the inspecting team and their salaries. It appears that Angus Smith had been appointed as Inspector early in 1864, although the papers concerning his appointment have not survived. It is unlikely that this was an open competitive process; it was rather a recommended appointment. Smith was a strong candidate for the post given his investigations into air quality, his work for parliamentary commissions and his commitment to environmental improvement. However, it is more than likely that the Board of

Trade sought some independent review. In that case, probably one of the first to be approached was Lyon Playfair. Playfair had played an important part alongside Prince Albert in the planning for the Great Exhibition of 1851 and became an adviser on scientific matters to the government without having the official title. From Playfair's and Angus Smith's work together at the Royal Manchester Institution and their collaborative investigations for the report on sanitary conditions in industrial Lancashire as part of the Royal Commission on the Health of Towns undertaken by Edwin Chadwick, Angus Smith's credentials and commitment were evident. Given the good working relationship between the two, Playfair would readily have supported Angus Smith's appointment to the Inspector's post.

What of the rest of the inspection team? Besides the Inspector's post, the Board of Trade, no doubt with Angus Smith's support, felt that four sub-inspectors were essential if the different regions of the country were to be adequately covered, given the overall number of alkali works and the number of visits that might prove necessary to ensure that the works condensed as much acid gas as possible, and certainly met the 95 per cent minimum limit. There was then the thorny difficulty of salaries, and confrontation with the Treasury was almost certain. The Board of Trade was responsible for the railway inspectors, whose salary was £1,000, but decided to seek advice from the Home Office, which was responsible for the mine inspectors and the factory inspectors. These posts were felt to be closer in operation and responsibility to the new posts, and might provide a better comparator for their discussions with the Treasury. In fact the responsibilities and powers of the inspectors for the Alkali Inspectorate were greater than those for the other inspectors: while the work of the latter was largely advisory, the alkali inspectors had legal powers and were able to prosecute in the courts.

On behalf of the Board of Trade, T.H. Farrer requested approval to appoint the Inspector at £1,000 and the four sub-inspectors at £500, but this request was refused by the Treasury, which nearly always challenged any comparative approach in such negotiations:

> having regard to the number and importance of the factories in the U.K. and the responsibility devolving on the inspectors involving moral as well as physical considerations ... would constitute the higher class of inspectors.[1]

The Treasury proposed reducing the salary of the Inspector to £700 and that of the sub-inspectors to £350, and cutting the overall staff by half. This response was not totally unexpected but it placed the Board of Trade in a major dilemma. The inspectorate team was implementing sensitive legislation in which a fundamental shift in the relationship between government and industry was occurring: a change from the *laissez-faire* approach that had served industry and government so well up to this time to the interventionist approach that was felt to be essential to

---

[1]   Letter dated 2 September 1863 from the Office of the Committee of the Privy Council for Trade to the Secretary of the Treasury. NA: (Ref: MH 16/1).

control the pollution from alkali works. With some manufacturers likely to offer resistance rather than cooperation, it was essential that inspectors were well trained scientifically, had experience of working in the chemical industry and would play independent roles. There was also the status attached to the Inspectorate; this might be undermined by a reduction in salary compared with other inspectors. In its reply the Board of Trade addressed the sensitivity of the work:

> having regard to the novel and delicate nature of the duties which they will have to perform ... should possess an exclusively official character and be protected from all suspicion of local or personal influence.[2]

In the end a compromise was agreed, with the Inspector's salary at £700 and four sub-inspectors at £400; the Inspector's post was regarded as part time to enable Angus Smith to continue with his chemical consultant work in Manchester. Interestingly, it was never made clear whether the part-time designation was genuinely to allow Angus Smith to pursue his consultancy work or was a convenient cover for justifying his reduced salary. Angus Smith certainly never saw the post as part time because he always carried a very heavy workload while working with the Inspectorate.

By February 1864 the four sub-inspectors were appointed: Alfred Fletcher based in Liverpool for the No. 1 (Western Division) area covering Lancashire, Cheshire and Flintshire; John Hobson in Manchester for the No. 2 (Middle Division) area covering East Lancashire, Birmingham, Yorkshire and London; Brereton Todd in Newcastle for the No. 3 (Eastern Division) area covering the Tyne, Middlesbrough and Seaham; Charles Blatherwick in Glasgow for the No. 4 (Northern Division) area covering all Scotland and Ireland. At the time of their appointment the *Chemical News* reported that their names were 'entirely unknown to us'.[3] Unfortunately, little is now known of the sub-inspectors except for Alfred Fletcher and John Hobson. Fletcher had studied chemistry and mathematics at University College London, for which he was awarded the gold medal and then pursued an interest in dyestuffs, becoming involved in a protracted patent dispute. He was a member of the Institute of Chemistry and the Chemical Society, being elected a Fellow in 1867. We know only that John Hobson was elected a Fellow of the Chemical Society in 1858. It seems probable that all the inspectors were at least well qualified in chemistry.

With the inspection team in place, the next stage was to draw up the inspection procedure. Looking back on his first report as Inspector, Angus Smith was able to record:

> The inspection demanded was of a kind entirely new: there was no well-recognized method of performing the work, and although we had succeeded

---

2   Ibid.
3   *Chemical News*, 69 (1884), p. 222.

very well in preliminary trials, many other plans were proposed, and the task of examining and deciding promised to be great; gradually, however, the process became simpler and the results produced confidence.[4]

Before finalizing the inspection procedure, the inspectors had to review the acid towers (or condensers, as they were often called) being used at all the works to understand their mode of working, whether they were working as effectively as possible and how their performance could be improved to meet at least the 5 per cent limit on escapes.

Acid Tower Improvements

It must have been a major relief to Lord Derby and other members of the 1862 Select Committee when William Gossage gave evidence and provided details of his tower, later known as the acid tower, for condensing the hydrogen chloride gas so effectively. However, Gossage must have felt very disappointed that the tower had not been more widely adopted in the period leading up to the Select Committee.[5] There are several reasons why the acid tower was not adopted earlier. Besides a lack of understanding of the principles behind the working of the tower, discussed in Chapter 5, another reason was the absence of a use for a major part of the large quantity of hydrochloric acid produced in the acid tower.

By the time of the Select Committee in 1862 most manufacturers were persuaded of the benefits of acid towers. A typical acid tower of the early 1860s is shown in Figure 6.1 and constructional evidence has come from archaeological surveys as well as documentary sources.

The towers varied in shape and size, depending on the conditions prevailing at the works, such as temperature of the furnace (saltcake pan) gas, the amount of gas to be condensed in each batch, and the required concentration of the resulting acid (this became an important consideration with later developments). They were constructed of brick or stone flags, which were cemented and bolted together, often with cladding on the outside. The bases of these towers often occupied up to 64 square feet of ground space, and at Hutchinson No. 1 works on Spike Island, Widnes, the foundations were made on bad ground, necessitating a raft of wooden beams on which Yorkshire stone slabs rested. The towers were packed with coke or acid-resistant brick. The coke towers often presented considerable operating difficulties. As Watt and Richardson point out,

---

[4]   *First Report of the Inspector*, p. 35.

[5]   It is interesting to speculate on the possible outcome of the 1863 Select Committee had Gossage not invented his tower. With the industry's important economic contribution, the alkali works could be grouped together in designated geographical areas where the acid gas would do little damage. Closing the works completely was probably not an option.

Figure 6.1   Acid tower at John Hutchinson's works in 1860s
Georg Lunge, *Sulphuric Acid and Alkali*, vol. 2 (London, 1886),
p. 252

If the coke is too tightly packed or the pieces are too large, there is not enough
surface opposed to the gas: the layers easily collapse, form compact masses, and
allow the gas to pass more easily through other parts of the sectional area. In this
case a large portion of the tower remains inactive, and consequently the really
efficient part of some very large condensers is very small. Exactly the same
thing happens when the coke is packed too tightly.[6]

The acid tower [in Figure 6.1] shows the use of acid resistant bricks, probably
'Obsidianite' bricks from Charles Davison and Co. Ltd. of Ewloe (near Chester).
In some high towers a grating or dome was built to throw the pressure downwards
and so keep it off the sides, and also to lessen the pressure of the packing on the
lower parts. The cover consisted of several stone flags with joints of red lead,

---

[6]   T. Richardson and H. Watt, *Chemical Technology* (London, 1863), vol. 1 (part 3),
pp. 307–8.

but allowing access for the water spray from the cistern above. The total height varied from 5 to 120 feet.[7]

A letter to Angus Smith from a manufacturer reveals the way many manufacturers were beginning to think about the advantage of building acid towers rather than paying financial compensation:

> The amount paid by him [the manufacturer] for the last four years has been 150*l* per annum for damages. The expense of alterations so as to bring condensation within the requirements of the Act has been 300*l*, so that the compensation in two years is sufficient to remove the grievance so far. Considering the sum of 300*l* as capital, condensation within the Act becomes a saving of expense.[8]

Manufacturers were gradually becoming aware that any loss of acid gas meant a loss of chlorine, which in turn meant a loss of income from bleaching-powder sales. But Angus Smith was reassured by the manufacturer realizing that this was only a preliminary stage and 'there is generally a desire to proceed further'.[9]

Possession of such towers was no automatic guarantee that they would be operated effectively, for close control by the works manager, foreman and other workers was vital if the towers were to condense all the acid gas effectively on a continuous basis. Since pay was linked to quantity of chemicals produced, there were considerable pressures for workers to take short cuts when the plant was not under close scrutiny, resulting in inefficient plant operation or, worse still, release of acid gas. There were also the nights when, with less management scrutiny and under cover of darkness, many of the serious releases of acid gas occurred. These episodes became serious challenges for the inspectors when they began setting out their monitoring procedures.

Modifications were made to acid towers to ensure their continued efficient working. One significant change from Gossage's original tower was the counter-current of the gas and water – water from the top and gas from the bottom – and this arrangement resulted in greater efficiency. The tower was placed between the saltcake furnace (pan) and the chimney, which was now able to perform its proper function by drawing the heavy gas through the towers. The draught had to be carefully controlled to ensure that the gas had sufficient time in contact with the film of water to allow absorption. However, when the conditions were carefully controlled, operation of the acid towers became very efficient.[10]

---

[7]   Peter Reed, 'Acid Towers and Weldon Stills in Leblanc Widnes', *Journal of the North Western Society for Industrial Archaeology and History*, 2 (1977), p. 4.

[8]   *First Report of the Inspector*, p. 79.

[9]   Ibid.

[10]   Ibid.

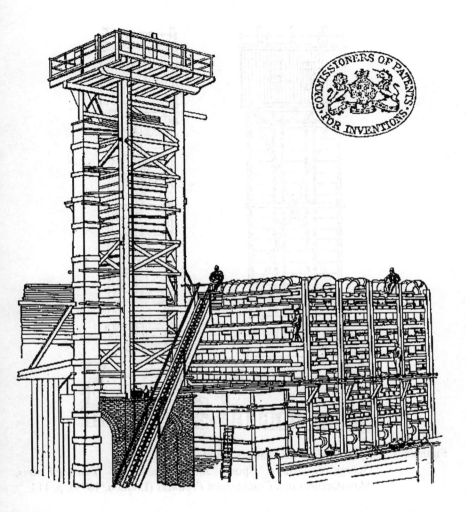

Figure 6.2   Cast-iron pipes for cooling gases before condensation in acid tower
George G. André, *Spon's Encyclopaedia of Industrial Arts,
Manufactures and Commercial Products* (London, 1879), p. 143

Efficient operation of the acid towers was taken up in detail during the examination
of scientific and manufacturing witnesses, and the Select Committee considered
the best way of achieving this outcome. The acid tower had to work in a robust
and reliable manner if legislation were to be built around its adoption. Following
Gossage's invention of the acid tower in 1836, several modifications were made
to improve efficiency with the increasing quantities of acid gas. One particular
concern was the temperature of the gas before it entered the acid tower; generally,
an increase in temperature of a gas will reduce its solubility in water. For Angus

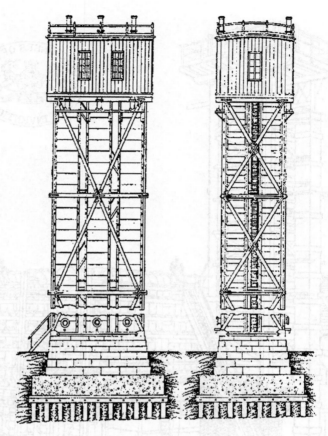

Figure 6.3   System of six acid towers coupled together
              George G. André, Spon's *Encyclopaedia of Industrial Arts,*
              Manufactures and Commercial Products (London, 1879), p. 114

Smith, cooling the gas was 'the key of every good condensation'.[11] As discussed
earlier, during the first stage of the Leblanc process when salt and sulphuric acid
are reacted together, there are two steps in the operation: in the second step, when
the pasty mass is roasted, large quantities of very hot acid gas are produced. The
hot acid gas could not be passed directly to the acid tower but must first be cooled.
The gas is at a temperature approaching 1,100 °C on leaving the roaster and must
be nearer 300 °C before entering the acid tower. Cooling was achieved by passing
the hot gas through a series of cast-iron pipes or brick flues of sufficient length to

---

[11]   G. Lunge, *A Theoretical and Practical Treatise on the Manufacture of Sulphuric*
Acid and Alkali (London, 1879), vol. 1, p. 191.

ensure that the gas was cooled to about 300 °C (see Figure 6.2).[12] The cast-iron pipes were supported off the ground to allow a free current of air to circulate over them. The acid towers were usually coupled together. Figure 6.3 gives two views of a system of six towers. This arrangement had a number of advantages. Additional strength was achieved by the exterior bracing and by the enlarged foundations; there was always concern about subsidence and the possible cracks it might cause in the fabric of the structure, resulting in escape of gas. Coupling of towers also ensured better use of space, in preference to having a number of individual towers scattered around the works. A further advantage was that it allowed the towers to be worked together. Complete condensation nearly always required passage of the gas through more than one tower, no matter how carefully the conditions within the tower were adjusted. Also, it was possible to pump the hydrochloric acid from the bottom of one tower to the top of the next to achieve the required concentration of acid. This was important when the acid was used to generate chlorine for bleaching powder.

The acid towers became characteristic landmarks at chemical works. For the alkali industry the towers provided the means for the industry to continue production by preventing release of the acid gas and enabling it to be reused for manufacture of another commercial product. Their function was replicated in other chemical works where there was a different function; often they acted as 'scrubbers'. While the towers were often known to collapse, the foundations with their Yorkshire stone flags remained undisturbed.[13]

Automating Aspirators

When reviewing possible inspection procedures, Angus Smith had

> dismissed as unscientific and unsystematic the suggestion that some people were sufficiently skilled as observers to enable them to differentiate between a 5% escape of acid gas and a 6% escape; instead he saw the need for chemical analysis to be rigorously applied and, therefore, the inspectors needed to be fully conversant with the methodology and application of chemical analysis.[14]

This approach was especially important when assembling evidence for any legal prosecutions.

---

[12] George André, *Spon's Encyclopaedia of the Industrial Arts, Manufactures and Commercial Product* (London, 1879), vol. 1, p. 143.

[13] When the land reclamation schemes were undertaken in the late 1970s and early 1980s, most of the foundations were left *in situ* because of their extreme weight. They remain to entice archaeologists in generations to come.

[14] Reed, 'Angus Smith and the Alkali Inspectorate', p. 159.

One of the first requirements was an accurate aspirator in which to collect the gases under test. Initially the inspectors used a box of one cubic foot capacity, but when filled with water this became heavy and was cumbersome to use within the flues from where the gases were collected. Aspirators were liable to damage during transit, particularly during train journeys. As the respect and cooperation between inspectors and manufacturers grew, the bulky aspirators were allowed to remain *in situ* within the plant at the alkali works.

Angus Smith also made available a more flexible design of aspirator based on one he had used when collecting air from towns and mines during his various studies. He described it thus:

> it is a bag of cylindrical form stretched out with hoops at intervals. There is a wide opening through which the air can be rapidly passed so as to empty the aspirator; ... A weight is attached to the bottom so that it may draw sufficiently, but the speed is regulated by the tap at the entrance tube. It can be made to draw the gas at the rate of one bubble in some seconds, or in a rapid stream which flows constantly and regularly. The amount of gas may also be measured with considerable and as a rule abundant accuracy. It has only one fault, the flexible material, which is caoutchouc cloth, is apt after a while to become imperfectly tight.[15]

The gas collected was expelled into a bottle of the test solution – in the case of acid gas, into a standard solution of silver nitrate.[16] Further volumes of air were similarly passed through the test solution.

The demand for more accurate and reliable aspirators attracted the attention of prominent instrument makers. One of these was John Dancer, who came from a family of instrument makers – Dancer's father was principal assistant to Edward Troughton, a distinguished instrument maker famed for his astronomical instruments.[17] From 1841 Dancer was based in Manchester and it is probably through their membership of the Manchester Literary and Philosophical Society that Dancer met Angus Smith and subsequently discussed the design of aspirators. Dancer in due course designed what became known as the 'swivel aspirator' (see Figure 6.4). This type of aspirator consisted of

---

[15]   Ibid., p. 41.

[16]   By using a standard solution (a solution of known concentration) of silver nitrate and measuring the amount of silver chloride precipitated, it is possible to calculate how much acid gas has passed through in the measured interval of time.

[17]   Edward Troughton replaced all the antiquated astronomical instruments at the Greenwich observatory. See Anita McConnell, 'Entry for Edward Troughton', *ODNB*, and also W. Skempton and J. Brown, 'John and Edward Troughton', *Notes and Records of the Royal Society*, 27 (1972–73), pp. 233–62.

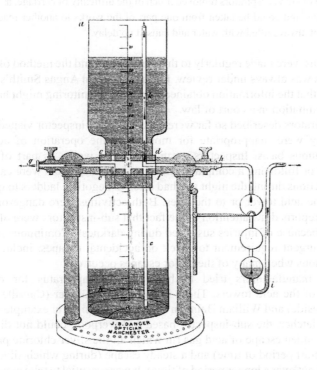

Figure 6.4   Swivel aspirator made by J.B. Dancer for Robert Angus Smith
                First Report of the Inspector appointed under the Alkali Act, P.P.
                1865 (3460), p. xx

two graduated vessels, one inverted with respect to the other, mounted on a brass
frame. The vessel filled with liquid was uppermost and as the liquid passed to
the lower vessel the air (or gas) under test was drawn into the upper vessel and
the air (or gas) originally in the apparatus was expelled. As the sample entered
the apparatus it passed through the test solution held in glass bulbs. The vessels
were graduated into parts of a cubic foot which enable the volume of the sample
to be read off accurately.[18]

It was much easier to use than earlier aspirators, as Smith reported:

> The use of swivel aspirators removed much of the difficulty of carriage, as it was
> not heavy and could be taken from one part of the works to another readily. It
> was kept always filled with water and caused no delay.[19]

Modifications were made regularly to these aspirators and the method of analysing gas samples was always under review, for uppermost in Angus Smith's mind was the thought that the information obtained during the monitoring might have to bear intense examination in a court of law.

The aspirators described so far were used when the inspector visited the alkali works. They were inappropriate for monitoring the operation of acid towers on a continuous basis. Inspectors normally visited works as part of a regular programme or following a complaint. On occasions inspectors were called out to make inspections during the night and had often to negotiate ladders to gain access to part of the acid tower or to the flues. Both activities were dangerous and the inspectors' reports draw attention to the fact that sub-inspectors were often absent from work because of injuries sustained during darkness. Continuous monitoring became an urgent requirement to check on accidental escapes, including night-time operations when many of the major escapes occurred.

Several manufacturers tried to build suitable apparatus for continuous monitoring of the acid towers. These included John Glover (Carville Chemical Works, Tyneside) and William Balmain in St Helens. An early example developed by Alfred Fletcher, the sub-inspector based in Liverpool, could not differentiate between a sudden escape of acid gas (in which all the silver chloride precipitated out over a short period of time) and a steady escape (during which silver chloride precipitated out over a longer period of time). It was essential to take measurements over distinct intervals of time, say between one and six hours, while maintaining the monitoring operation over several days. This requirement led to the compound self-acting apparatus (see Figure 6.5) developed by Fletcher that ingeniously depended for motive power on the draught of the chimney. A fan (about 2 inches in diameter) was set into the flue and the air drawn into the flue caused the fan to revolve, which, through an endless screw, operated a small bellows pump made of vulcanized rubber. The bellows drew a stream of air from the chimney or flue through a bottle containing the test solution (in the case of acid gas, silver nitrate). The time set for the gas to pass through the first bottle was adjusted between 3 and 6 hours, after which it passed through the second bottle and so on. The total volume of gas passing through the apparatus was recorded on a gas meter.

A particularly elegant feature of the apparatus was its ability to record the time at which any one bottle was receiving gas:

---

[19]   Ibid, p. 42.

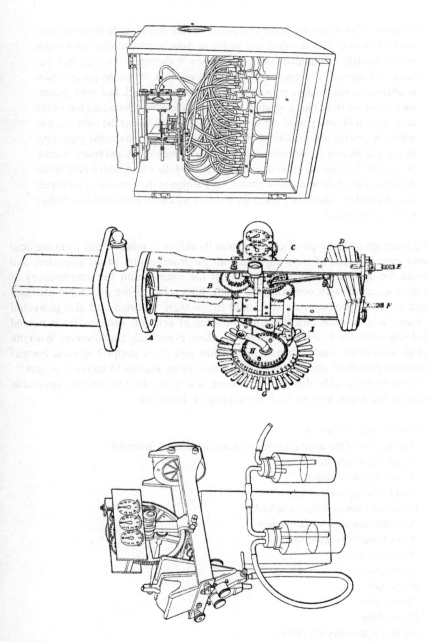

Figure 6.5   A compound self-acting aspirator, as developed by Alfred Fletcher
George G. André, *Spon's Encyclopaedia of Industrial Arts,
Manufactures and Commercial Products* (London, 1879), pp. 132–
3. The British Library Board

A ribbon of photographic paper enclosed in a dark box is made to unroll from one reel on to another at a slow rate, and in so doing to pass another slit through which daylight is admitted. The paper passes at the rate of half an inch per hour, and becomes darkened as it passes the opening. When the paper ribbon is afterwards removed, it presents a series of alternate dark and light spaces, each about six inches long, shading off one into the other. The centre line of the dark space will have passed the slit at noon, the centre line of the light space at midnight, and the intermediate points at intervening times, so that the paper may be marked out and divided into spaces corresponding to days and hours. A mark is also imprinted on the paper at each revolution of the wheel, which determines the connexion of the successive bottles; the position of these marks on the paper, now divided out into hours, gives the times at which the corresponding bottles were in operation.[20]

The advantage of this type of aspirator was its ability to monitor gas from the acid towers over a period of nine days without the intervention of an inspector, and because the apparatus was sealed, without the intervention (or interference) of anyone concerned with operating the plant. This relieved the inevitable pressure placed on an inspector when accidental losses were reported and also protected the manufacturer from unsubstantiated claims of acid-gas escape. But it passed added responsibility to those operating the plant, principally the foreman. With the level of production linked to pay, production was often stepped up (and normal procedures curtailed) during hours of darkness in an attempt to increase wages.[21]

While reviewing the different aspirators, it is interesting to note the apparatus carried by the inspectors for their monitoring. It included:

Fine balance and weights
A collection of the usual chemical reagents in solution in bottles
A cubic foot aspirator
A swivel glass aspirator
Simple fan aspirator or tell-tale
Compound fan aspirator or tell-tale
A flexible aspirator without water
A finger pump as aspirator
A pressure gauge
Anemometer
A foot rule
Hydrometer
Glass globes
Air pump to empty the globes

[20]   Ibid., p. 53.
[21]   Robert H. Sherrard, 'The White Slaves of England', *Pearson's Magazine*, 2 (1896), pp. 48–55.

Adsorption tubes and stands
Beaker glasses and globes
Burner for gas
And a few other things required in small laboratory experiments.[22]

Such a quantity of equipment was a bulky and cumbersome load to take out on each visit. As the relationship between inspector and manufacturer became more cooperative, it was probably possible and even advantageous for the inspector to leave some of the bulkier equipment on site. Many manufacturers, in response to the legal limit of 95 per cent for acid-gas emissions set by the 1863 Act, appointed their own chemists to carry out regular analyses of the gases passing down the flues and released from the chimneys. As historians Colin Russell, Noel Coley and Gerrylynn Roberts have pointed out, it was the Alkali Act 1863 and its greater dependence on analytical chemistry that advanced the professionalization of chemists and in particular the foundation of the Institute of Chemistry in 1877.[23]

Inspectors as Peripatetic Consultants

With the Select Committee hearings, the continuing court cases claiming damages and the subsequent approval of the Alkali Act in 1863, Angus Smith could have reasonably expected manufacturers not only to be aware of their responsibilities regarding acid-gas emissions from their works but to have made sure that their works met the legal limits. However, as he was to find:

> Sometimes works have been standing still in order that improvements might be made. And sometimes it was advisable to consult as to the best mode of making the improvement. It has been necessary to be very careful not to give such advice as might appear to be an interference; we have no power to decide on the mode of producing the desired result. At the same time, it so frequently happens that the necessary improvement is self-evident to those who are experienced, that any hesitation to advise would be unfair and indeed simple pedantry. This is more especially seen when the manufacturer himself has not at all thought on the subject, and is obliged either to seek advice or to remain inactive.[24]

Smith was diplomatically pointing out that for many of the alkali works the expertise to make the necessary operational changes was missing. How had this situation come about?

Most owners of alkali works were entrepreneurs who had perceived a good business opportunity, just as James Muspratt had done in moving from Dublin to

---

22 *First Report of the Inspector*, pp. 60–61.
23 Russell, *Chemists by Profession*, pp. 101–3.
24 *First Report of the Inspector*, p. 58.

Liverpool in 1822 to meet the demands for alkali of the Merseyside soapboilers, and left the day-to-day operation of the works in the hands of a foreman whose only source of knowledge was that gained on the job. Consultant chemists were brought in to make any technical modifications or innovations to the process, though sometimes this expertise was available through a family connection. This became increasingly important from the 1830s as business owners recognized the value of education, and for chemical entrepreneurs this was chemical education and training provided by such leading figures as Thomas Graham (Glasgow) and Justus von Liebig (Giessen and Munich). Later this group expanded to include A. Wilhelm Hofmann (Royal College of Chemistry, London) and Robert Bunsen (Marburg and Heidelberg).[25] Under the influence of Liebig, much of the training concentrated on analytical chemistry, and while such knowledge and expertise was useful in analysing raw materials and chemical products, it had very limited application to modifying chemical plant or changing the operation of plant. This work was left to the few owner–entrepreneurs with an inventive turn of mind (such as William Gossage) or to consultants (such as Ludwig Mond and Walter Weldon). As Georg Lunge reported, professionally trained chemists were not widely employed in alkali works, or other chemical works, until the 1870s. The first degree courses in Britain covering pure and applied chemistry with some chemical engineering were not offered until the later 1890s.

When Angus Smith took over as Inspector of the Alkali Inspectorate in 1864 he found that managers of alkali works possessed little understanding of chemistry and chemical technology. If alkali works were to meet the legal limit set for the emission of acid gas, Angus Smith and his inspection team would have to take a more proactive role than may have been envisaged originally. In the longer term this 'peripatetic consultant' role helped the owner–entrepreneurs to build a better relationship with the inspectors: their initial caution gradually gave way to a more accepting and cooperative relationship that benefited both parties.[26] The inspectors achieved levels of condensation that were considerably above the minimum laid down by the legislation, while most owner–entrepreneurs kept their works free of claims for financial damages. But there was a tightrope to walk for the inspectors arising from the nature of the work undertaken and the benefits that might accrue. Not only was the plant operating more efficiently, but the works was

---

[25]   For a wider discussion of these issues, see J.F. Donnelly, 'Representations of Applied Science: Academies and Chemical Industry in Late Nineteenth-Century England', *Social Studies in Science*, 16 (1986), pp. 195–234, and J.F. Donnelly, 'Consultants, Managers, Testing Slaves: Changing Roles for Chemists in the British Alkali Industry, 1850–1920', *Technology and Culture*, 35 (1994), pp. 100–128.

[26]   This role is explored further in Vladimir Jankovic in a paper 'Bad Air in Courts of Law: R.A. Smith on the Nature of Sanitary Evidence', at the conference 'Good Airs and Bad: Historical Perspectives on the Atmosphere in Relation to Health and Medicine', Wellcome Unit for the History of Medicine, University of East Anglia, 10–11 November 2000.

very probably returning a better commercial profit. Such accusations beleaguered Angus Smith and the Alkali Inspectorate whenever the role of the inspectors was under review, and especially when there was any expression of a better working relationship with the manufacturers after initial frosty relationships. Nevertheless, this relationship proved to have an important bearing on future legislation and on its implementation, employing the 'best practicable means', approach as we shall see in Chapter 7.

Inspector's Annual Reports to Parliament

One of the principal stipulations in the Alkali Act (1863) was the requirement for the Inspector to provide an annual report to Parliament, although the brief for its contents was vague:

> The Inspector shall, on or before the First Day of March in every year, make a report to the Board of Trade of his Proceedings during the preceding Year, and a copy of such Report shall be laid before both Houses of Parliament.[27]

Angus Smith was therefore left to decide for himself what to include. In the first few reports he concentrated on the state of the alkali industry, its ability to condense at least 95 per cent of the acid gas, as well as details of the different condensers employed to do this. The first report includes considerable details of the analytical techniques used to measure the quantity of acid gas in different part of the plant and outlines the development of the different aspirators used by the inspectors to assist with collecting gas samples for analysis. This included aspirators that could take samples on a regular basis over many days. These sections were probably included to reassure Angus Smith's colleagues in the Board of Trade and in Parliament of the seriousness of the Alkali Inspectorate's work and the thoroughness with which Angus Smith and his team of inspectors ensured that the terms of the Act were met by every alkali works.

But there are other interesting sections in the reports. All the alkali works operating the Leblanc process in the UK are listed, but when the operation of any particular works is discussed in technical detail the name of the works is omitted and the works is referred to only by its registration number (all works were required to register).[28] This ensured that the Alkali Inspectorate had a full list of all the alkali works and each inspector could plan the schedule of visits within his division. The registration of works and assignment of a registration

---

[27]   *Alkali Act 1863*, P.P. 1863 (135), p. 3.

[28]   The original register has survived and is now held by the Health and Safety Executive. By ascertaining the works number from the register it is possible to obtain a considerable quantity of information about an individual works.

number allowed the inspectors to keep commercially sensitive information about individual works concealed from competitors.

Part of this sensitive information comprises detailed technical specifications on the extent of plant at each works and its production capacity. At the end of the first report Angus Smith asks rhetorically if he has provided too much information and whether it is useful for those reading the report. It probably made little sense to the parliamentarians, who were much more concerned about the headline statistic, the overall rate of condensation of the acid gas and whether it was higher than 95 per cent. Fortunately, even in his first report in 1864, Angus Smith was able to conclude that condensation of acid gas was 98.72 per cent; 99.11 per cent by the second report (1865) and 99.27 per cent by the third report (1866).

There is another reason why Angus Smith was collecting this prodigious quantity of technical information. From the 1830s a movement of statistical societies had began that collected data as a means of improving social policy, just as experimental data had been used to further progress in the natural sciences. In 1833 the Manchester Statistical Society was founded with the aim 'to assist in the promoting the progress of social improvement in the manufacturing population by which they are surrounded ... and to the promotion of statistical inquiries'.[29] At a national level, the best example of this endeavour was the work done by William Farr in the General Register Office from 1839, with the statistical information collected on deaths from contagious diseases as the First Annual Report of the Registrar-General:

> Diseases are more easily prevented than cured, and the first step to their prevention is the discovery of their exciting causes. The registry will show the agency of their causes by numerical facts, and measure the intensity of their influence.[30]

Angus Smith's annual reports as Inspector also adopt this statistical approach, collating a vast amount of detailed information about each alkali works and many facets of their operation. The first report includes tabulations with the headings, 'Some Statistics showing the General Magnitude of the Works and the Mode of Working'[31] and 'Amounts of Condensation shown by the Percentage and Actual Amount of HCl [Hydrogen Chloride] Escaping'.[32] While the second report goes

---

[29]   *First Annual Report*, Manchester Statistical Society, Manchester Central Library. See also Thomas S. Ashton, *Economic and Social Investigations in Manchester 1833–1933* (Brighton, 1977), p. 13.

[30]   *First Annual Report of the Registrar-General*, *Appendix (P)*, P.P. 1839 (187), pp. 64–5. For the benefit of this statistical approach in preventing disease, see Pamela K. Gilbert, *Mapping the Victorian Social Body* (Albany, NY, 2004), and in particular, Part 2 – 'Mapping Disease in the Metropole', pp. 27–82.

[31]   *First Report of the Inspector*, p. 27.

[32]   Ibid., p. 10.

into even greater detail, with 'Particulars of the Condensation',[33] 'Some Statistics Showing the Mode of Working and the Amount Done'[34] and 'Particulars Relating to the Cooling and Condensing Apparatus'.[35] The third report includes 'Temperature of Inlet and Outlet Flues'.[36] Figure 6.6 illustrates some of the tabulation headings.

Tabulation of Technical Data by Alkali Administration

Some Statistics Showing the General Magnitude, etc.[1]

Register Number.	Tons. of Salt decomposed per week.	Number of decomposing pans at work.	Cwts. of Salt per Charge.	Number of charges in 24 hours.	Cwts. of salt Decomposed In 24 hours.	Number of open Salt Cake Furnaces.	Number of close Salt Cake Furnaces.

Particulars of Condensers[2]

Register Number	Gas entering Condenser grains of HCl per cub. ft.	Gas leaving Condenser grains of HCl per cub. ft.	Gas entering Condenser grammes per cub. metre.	Gas leaving Condenser grammes per cub. Metre.	Pan per cent. of escape	Total per cent. of Escape	Cwts. of HCl per day.

Particulars Relating to the Condensers[3]

Register Number	Cooling Space in Pipes, Tanks etc.	No. of Towers	Dimensions		Material used for filling.	Cubic Contents in Feet.
			Sectional Area	Height		

[1]   *First Report of the Inspector appointed under the Alkali Act*, P.P. 1865 (3460), p. xx.

[2]   *Second Report of the Inspector appointed under the Alkali Act*, P.P. 1866 (2701) p. 10.

[3]   *Third Report of the Inspector appointed under the Alkali Act*, P.P. 1867 (3792), p. 8.

Figure 6.6   Headings for technical data collated by Alkali Inspectorate Taken from Annual Reports of the Inspector appointed under the Alkali Act

---

[33]   *Second Report of the Inspector appointed under the Alkali Act*, P.P. 1866 (3701), p. 10.

[34]   Ibid., p. 12.

[35]   Ibid., p. 14.

[36]   *Third Report of the Inspector appointed under the Alkali Act*, P.P. 1867 (3792), p. 38.

It is difficult to understand why all this information was useful in this case, except that it gave the inspectors a fuller picture of the different parts of the works and the way in which they worked. However, such detail was unlikely to reveal the best set-up or the most efficient working because each works was unique in its construction, composition of plant and its operation. If Angus Smith had not realized this at the outset, perhaps he appreciated the situation later, for the level of information detail decreased in later reports, where the emphasis focused more on industries and gases that should be put under regulation. It is interesting to note that Angus Smith used a similar approach with his major book on air quality, *Air and Rain: The Beginnings of a Chemical Climatology*, published in 1872, as was discussed in Chapter 4.

Angus Smith and his successors saw the annual reports as an opportunity to draw the attention of senior civil servants in the Board of Trade (and later in the Local Government Board) to shortcomings in the existing legislation, whether they were changes to the legal framework to assist the Alkali Inspectorate in its objective of preventing harmful gases from chemical processes being released into the atmosphere or the necessity to bring other industries and gases under regulation where the introduction of new processes or working was proving detrimental to air quality. Later, as we shall see, Angus Smith worked with colleagues in other government departments to highlight the public health aspects of these gases and their impact on working and living conditions; by working in a concerted and coordinated manner Angus Smith (and his colleagues) felt that there would be a greater prospect of a successful outcome in Parliament.

Comparison with Regulation in France

It is interesting to compare the approach to regulation taken in France, where the Leblanc process originated. Angus Smith had visited France in 1855, before being appointed Inspector of the Alkali Inspectorate, with the purpose of investigating the approach taken there in respect of the regulation of different types of works. While his interest was primarily in chemical works, the regulations related to any works where there was the likelihood of smoke, noxious vapours or fire. He used his fourth report as Inspector in 1867 to bring together information then available about how the regulation of works was organized in France following a more recent visit there.

In France the regulation of chemical works was controlled by a decree issued by Napoleon in October 1810 that divided chemical works into three classes depending on their potential danger to the public, as follows:[37]

---

[37]   See Thomas Le Roux and Jean-Baptiste Fressoz, 'Protecting the factories and commodifying the environment: the great transformation of French pollution regulation, 1700–1840', in Geneviève Massard-Guilbaud and Stephen Mosley (eds), *Common Ground. Integrating the Social and Environmental in History* (Newcastle, 2011), pp. 340–66.

The first class will comprehend those which ought to be distant from all private dwellings.

The second, the manufactures and workshops which it is not necessary to separate rigorously from dwellings, but which, nevertheless, ought not to be allowed before ascertaining correctly that the operations will be carried on in such a manner as not to cause damage or inconvenience to the neighbourhood.

In the third class will be placed those establishments which may remain without inconvenience before habitations, but ought to be submitted to the surveillance of the police.[38]

The allocation of a particular works to a class relied on a thorough understanding of the operations carried out at works of this type and their dangers to local inhabitants. In 1810, when the first lists were compiled, they were very short but as the chemical industry expanded in its sophistication and under regular review of the authorities, they grew longer. By 1867 they were extensive. Highlighting a few of the entries in each class will clarify the differentiation. Class 1 includes public slaughter-houses, menageries, arsenic works, soda works where the acid was not condensed, works manufacturing fireworks or matches, glue works and tanneries. Class 2 includes soda works where the acid was condensed, bleaching works, india-rubber works, lime kilns, chlorine manufacture, blast furnaces and potteries. Class 3 works include breweries, brickworks, gold and silver beaters, manufacture of red lead and litharge, paper manufacture and soapworks.[39] The list of works in each class is quite long and the same type of works could be in more than one class, depending on the chemical pathway of its processes and whether noxious vapours were condensed or removed before they were released into the atmosphere.

Administration of the regulations was in the hands of the Préfecture de Département in the regions, while in Paris the Préfecture de Police took responsibility for the special procedures that operated because of the tight control required in such a densely populated area. A manufacturer wishing to build a works was required to seek the agreement of the Préfecture before work began. An 1866 report on the workings of the regulations reported serious non-compliance, both with the initial registration and with the operation of the works:

> To give an idea of the state of things, it may be mentioned that the statistics for Herault for 1859 show that 277 establishments existed before the degree of 1810, whilst 1,931 have a later date. Out of this number 1,342 act without authority.

---

[38] *Fourth Report of the Inspector under the Alkali Act 1863*, P.P. 1867–68 (3988), p. 89.
[39] Ibid., pp. 92–105.

Of the 589 provided with authorisations, 122 comply with the conditions, whilst 413 evade them. 54 are without conditions.[40]

Enforcement was haphazard and lax, and inspectors were not employed to police the regulations to ensure that they were adhered to. As a result of this report some of the regulations (and classes) were simplified and inspectors began to be employed. In Paris, procedures were drawn up separately in each arrondissement of the city to ensure stricter compliance.

Angus Smith was convinced that the British approach, with its adoption of a limit on emissions for hydrogen chloride of 5 per cent, was working for all concerned, but he appreciated that 'the object now to be desired is a fixed point for other gases'.[41] But he was also very realistic about the need for further investigation:

Experience has shown numerous practical difficulties, and enquiry must be made on some other points, and numerous experiments collected, before we can expect to make a proposal which shall avoid that indefiniteness which is the fault of so much of the sanitary legislation in this as in other countries.[42]

Improvement in regulation and the setting of limits for other gases were actually to come from Angus Smith's experience of enforcing the existing legislation as much as with further experiments and investigations.

Indefinite Extension of Legislation in 1868

The 1863 legislation was due to terminate on 1 July 1868. Whether Angus Smith was asked by the Board of Trade to produce a final report is not known: if he was, the report has not survived. However, with the imminent termination Angus Smith used his fourth report as Inspector at the end of 1867 to draw some conclusions on success to date and to look ahead to expanding the terms of the legislation. The Alkali Inspectorate had seen the number of alkali works under regulation grow from 85 to 114, and the amount of salt decomposed rise to 371,000 tons annually. He openly recognized that escapes remained at about 2 per cent; these represented measured escapes from works and not those that were accidental or through malfunctioning of plant.

Angus Smith may have had prior indication of the legislation's extension because his conclusion pointed to the success to date and how other gases required regulation:

---

[40]   Ibid., p. 107. Hérault is a département in the Languedoc-Roussillon region of France.

[41]   Ibid.

[42]   Ibid., p. 116.

We may say that no attempt of any consequence has been made until the Alkali Act fixed on a number (5 per cent.) as the amount of permitted escape of muriatic acid. As we have this small point fixed, it is exceedingly important to keep the ground. I do not expect it to prove the best of all, but in the midst of uncertainty it is pleasant to have one certainty. The object now to be desired is a fixed point for other gases. I have had from the very beginning this object in view, and have hinted at it. I seemed to see a way to its accomplishment, but am unwilling to give any opinion until certain enquiries are made, and these I hope to accomplish.[43]

When the Alkali Act (1863) Perpetuation legislation came before Parliament it was approved on 4 June 1868, with section 19 of the original legislation repealed and two other sections removed.[44] The way was clear for Smith to build on what had been achieved to date and to consider other gases and industries that might be brought under regulation.

Working within the Local Government Board

Angus Smith had begun to find working within the Board of Trade quite frustrating. There seemed little understanding of the work of the Alkali Inspectorate and little support for its operations. The situation may have arisen in part because Angus Smith was always reluctant to have his work base in London and therefore at the heart of political and governmental machinations, but preferred to retain his home and laboratory in Manchester, a much more central location for work undertaken on behalf of the Alkali Inspectorate. The Board of Trade was in turn reluctant to administer a body imposing constraints on the working of industry as a result of the shift from a *laissez-faire* approach to an interventionist one, with inspectors having unlimited powers to enter alkali works, even though in reality the inspectors were acting as peripatetic consultants to industry. Perhaps they were even more cautious because Angus Smith in both his correspondence and in his annual reports as Inspector was highlighting cases of pollution caused by other industries and gases, while also proposing a more thorough measurement of the amount of acid gas released from alkali works. From his private consultancy work investigating the constituents of the atmosphere and how variations could influence health, Angus Smith could see strong links with his work for the Alkali Inspectorate. Perhaps the final straw in this testy relationship came in November 1868 when Angus Smith wrote to the Sir Louis Mallet, Permanent Secretary at the

---

[43]   *Fourth Report of the Inspector*, p. 115.

[44]   The sections removed were section 10 (salaries as may be determined by the Board of Trade, with the consent of the Commissioners of Her Majesty's Treasury) and section 17 (all penalties paid into the Exchequer). Strangely, the Treasury had been expected to block the removal of both!

Board of Trade, requesting a salary increase so that he could concentrate full time on the work of the Alkali Inspectorate. The request was turned down, probably without reference to the Treasury, which was nearly always the final arbiter in such matters. As Roy McLeod has concluded:

> However important the alkali works had become, the lack of any record of cooperation between the Board of Trade and the Inspectorate strongly suggests that though industry had by now recognized the State, the State had very little understanding of or appreciation for the scientific requirements of industry.[45]

By 1872 Angus Smith had advanced his ideas further about air quality under the theme of chemical climatology and on the link between composition of the air and health. The Public Health Act was passed in the same year and one of its clauses was the transfer of the Alkali Inspectorate to the newly created Local Government Board.[46] With the latter's responsibility for public health under the leadership of John Simon as Medical Officer, Angus Smith was more likely to find a sympathetic understanding for his proposals from the Permanent Secretary, John Lambert. Nevertheless, he would also need to be vigilant to ensure that the inspectors were retained as part of a centrally administered body rather than transfer to local authority control; this was one of the guiding principles of the Board with its responsibilities for Poor Law Administration and Public Health, among others. The increasing synergy between the work of the alkali inspectors and the public health remit of the Board, together with the prospect of working more closely with John Simon, may have prompted Angus Smith to make his views known to some parliamentarians and civil servants; however, if he did indeed make such representations, unsurprisingly no records have survived.

Widening Regulation in 1874

Now working in the Local Government Board, Angus Smith hoped for a better working relationship with John Lambert (Permanent Secretary) and with George Sclater-Booth (President) than had been the case in the Board of Trade. But at the early stages Angus Smith felt again that the Alkali Inspectorate was marginalized compared with the main departments of the Board. He continued to issue his annual report as Inspector and used every opportunity to indicate where the legislation had to change to reflect the newer challenges from other forms of pollution:

> Chemical works generally are greatly on the increase, and the power to repress escapes of gases does not increase with them ... The Alkali Act, which was excellent for a time and has done some good, is becoming less valuable daily.

---

45   MacLeod, 'Alkali Acts Administration', p. 94.
46   *Public Health Act 1872*, P.P. 1872 (261), p. 19.

When alkali works accumulate in one place they make even 1% of escape a great
evil ... However, I am merely repeating myself.[47]

Frustration was beginning to show, but Angus Smith was never deterred from
his principal responsibility to lead the Alkali Inspectorate. However, the ethos
within the Local Government Board gradually began to change for Angus
Smith. He started to develop a good working relationship with John Lambert,
who was known as a strict administrator and probably most preoccupied with his
conflict with John Simon, perhaps the most high-profile reformer in the public
health movement. Developing any useful relationships with other departments
was extremely difficult for Angus Smith while he retained his Manchester base,
and this remained the case with Simon. Roy McLeod has drawn attention to
Simon's supposed lack of interest in this branch of public health, as shown by
its omission from his major study, *English Sanitary Institutions*, but Simon made
some very pertinent and supportive statements while giving evidence to the Royal
Commission on Noxious Vapours in 1876, as we shall see in Chapter 7.[48]

It is impossible to gauge from the surviving papers how the relationship
between Angus Smith and Lambert blossomed, but perhaps through dogged
determination Lambert became convinced as early as 1874 of the need for changes
in the legislation, as Angus Smith had set out over many years. Lambert in turn
won over Sclater-Booth and the changes were incorporated into the Alkali Act
(1863) Amendment Act in 1874. The Act was progressive in nature: adapting
current clauses, omitting others and including new clauses, all in an attempt
to make the legislation relevant and effective in controlling the pollution from
chemical processes.

There were a number of important changes in the legislation:[49]

1. Included were works operating the Henderson process in which copper is
    extracted using salt with the release of hydrogen chloride gas.[50]
2. The percentage measure for acid gas was replaced by a volumetric measure
    that made sure that in 'each cubic foot of air escaping into the atmosphere
    there is not contained more than one-fifth part of a grain of muriatic
    acid'.[51] The inspectors had struggled to get accurate percentage releases
    because it necessitated knowing the total amount of acid gas produced as
    well as the amount escaping. The new approach would not only make the
    measurement easier, but also provide a more accurate assessment using
    methods of analytical chemistry honed by the inspectors over many years.
    Angus Smith was also conscious that even the current levels of escape

---

47   *Ninth Report of the Inspector appointed under the Alkali Act*, P.P. 1874 (C.815), p. 35.
48   John Simon, *English Sanitary Institutions* (London, 1890).
49   *Alkali Act (1863) Amendment Act 1874*, P.P. 1874 (99).
50   The Henderson process was patented in 1858. For details, see Haber, *The Chemical Industry*, p. 103.5[51] Alkali Act (1863) *Amendment Act 1874*, P.P. 1874 (99), p. 2.

could do great damage and needed to be reduced gradually – he proposed a yearly reduction of 0.01 of a grain, working closely with the manufacturers.

3. There was to be a greater responsibility placed on the manufacturers whereby 'the owners of every alkali works shall use the best practicable means, within a reasonable cost, of preventing the discharge into the atmosphere of all other noxious gases arising from such works, or of rendering such gases harmless when discharged'. This clause was to remain a bed-rock in later legislation when regulation was extended to other chemical processes.

4. Special rules made by the owner for the guidance of his workmen may extend to cover prevention of discharge of the noxious gases.

5. Noxious gases now included sulphuric acid (sulphur trioxide), sulphurous acid (sulphur dioxide) (except from the combustion of coal), nitric acid, or other offensive oxides of nitrogen, hydrogen sulphide and chlorine. This is a much more extensive list of pollutants than in earlier Acts. It is interesting to note the inclusion of hydrogen sulphide since this was a belated attempt to control the dumping of sulphur waste and its debilitating effect on both the natural and human environments. The position with coal smoke remained sensitive for the government because of the likely impact of any prohibition, but Angus Smith had begun to raise the issue in his annual reports even though such smoke was outside the terms of the legislation. He would continue (as would his successor) to take every opportunity to agitate for urgent action because of its blighting effect on health.

The legislation was to operate from 1 March 1875, and Angus Smith was delighted at last to have succeeded in updating the terms of the legislation:

> The Alkali Act of 1863, after 11 years of solitary struggle with noxious gases, receives assistance from the Act of 1874, and now begins in a small degree a more vigilant, and it is hoped a more useful life.[51]

Angus Smith could be satisfied for the time being, but the situation never stood still; other industries and new chemical processes produced new pollutants requiring regulation, and the public protests were to get more concerted. For how long would the Local Government Board offer its support to Angus Smith and his team of inspectors?

---

[51]   *Eleventh Report of the Inspector appointed under the Alkali Act*, P.P. 1876 (C. 1339), p. 3.

Chapter 7

Extension of Regulation from the 1870s and Increasing Health Concerns

Whereas Robert Angus Smith and his team of inspectors could point to high levels of condensation of hydrogen chloride gas by 1874, they also recognized ongoing occasions when, through accident or intentional neglect, large quantities of gas were released into the atmosphere. The only way to prevent these escapes was by encouraging owners, foremen and charge-hands operating the plant processes to act in a consistently responsible manner. Although Angus Smith and his team were never complacent, many landowners and their agents, as well a variety of campaigning groups, were always ready to undermine the overall progress when opportunities arose. This included lobbying the Local Government Board or presenting evidence to commissions set up to review the workings of current legislation. Accusations of lack of thoroughness were often targeted at Angus Smith and his team, and attention was increasingly drawn to other offending pollutants that urgently needed some form of regulation.

Survey of Unregulated Sectors and Pollutants from 1874

By the 1870s, Angus Smith and his colleagues were observing stark evidence of damage caused by chemical works other than those using the Leblanc process. Constant vigilance was required on the part of the inspection team; government departments and Parliament came under pressure to react quickly to new developments and circumstances regarding chemical pollution.

Sulphur Waste

The Leblanc process was responsible for a second pernicious pollutant, namely sulphur waste, and the accompanying emanation of the toxic gas hydrogen sulphide, with its characteristic 'bad egg' smell. The sulphur waste resulted from the third stage of the Leblanc process, when black ash was agitated with water (or lixiviated, as it was known in the alkali trade). The sodium carbonate dissolved in the water. The residue was the sulphur waste; this waste was also known by other names, including alkali waste, tank waste, vat waste and galligu (see Chapter 5).[1]

---

[1] Galligu was the name given to sulphur waste by the people of Widnes. It's a very appropriate onomatopoeic word to describe the evil-smelling black viscous material.

For every ton of soda produced, between one and a half and two tons of waste were created, with 15–20 per cent of sulphur in the waste.[2] By the 1870s, the annual quantity of waste across Britain was perhaps close to 500,000 tons. The waste contained up to 90 per cent of the sulphur in the sulphuric acid used in the first stage. By the late 1830s, attempts were being made to regenerate sulphur from waste and recycle it to produce more sulphuric acid. An analysis of alkali waste was completed by Muspratt and Dawson.[3]

Sulphur was an expensive commodity, and was mainly imported from Sicily, often with wide fluctuations in price.[4] To safeguard regular supplies at stable prices, several manufacturers purchased sulphur mines in Sicily, which proved problematic.[5] Manufacturers then looked to iron pyrites for their source of sulphur, but the problems concerning sulphur waste remained. This encouraged attempts to develop processes to reclaim the sulphur from the waste.

In 1837, William Gossage, inventor of the acid tower, patented a process for treating sulphur waste.[6] Gossage's approach was to treat the waste with carbonic acid (an aqueous solution of carbon dioxide), producing hydrogen sulphide. This was then burnt to form sulphur dioxide, which was fed back into the lead chamber for the production of further quantities of sulphuric acid. Gossage added a refinement: the hydrogen sulphide could be stored in a gasholder until it was needed, as a means of regulating its reuse. James Muspratt was sufficiently concerned about the sulphur problem to be convinced of the efficacy of Gossage's invention, and agreed to erect the necessary plant at his Newton-le-Willows works, at a projected cost of £500. Although the process worked well on a small scale, when scaled up it proved difficult to operate effectively, given the large quantities of waste. Gossage persevered for some time, but eventually had to abandon the process, leaving Muspratt with a bill reputed to have been £5,000 – a very considerable amount of money in the 1830s.[7]

---

[2]   C.T. Kingzett, *The History, Products and Processes of the Alkali Trade* (London, 1877), p. 133.

[3]   Ibid., p.134.

[4]   R.W. Rawson, ' On the Sulphur Trade of Sicily and the Commercial Relations with that Country and Great Britain,' *Journal of the Statistical Society*, 2 (1839), p. 449.

[5]   In 1834–35, James Muspratt and Charles Tennant worked together to purchase sulphur mines in Sicily, but in 1836 the king of Naples seized the mines. After intense diplomatic efforts, Lord Palmerston dispatched part of the British fleet to uphold British commercial interests, but the King of Naples remained resolute.

[6]   On 17 August 1837, William Gossage filed a patent (B.P. 7416/37) for treating sulphur waste and regenerating sulphur as sulphur dioxide.

[7]   See 'James Muspratt' in Fenwick Allen, *Some Founders*, p. 89.

Figure 7.1   William Gossage (1799–1877), chemist, manufacturer and inventor
              J. Fenwick Allen, *Some Founders of the Chemical Industry* (London,
              1906), facing page 1

Although such attempts were made to recycle the sulphur, the usual approach was to dump the waste on land surrounding the alkali works. With the increased levels of alkali production, Angus Smith reported in 1876, as part of his intermediate report on the work of the Alkali Inspectorate leading up to the Royal Commission on Noxious Vapours:

> The town of Widnes is very frequently, if not at all times, subjected to the influence of sulphuretted hydrogen [hydrogen sulphide] ... The tank waste, composed of sulphur and lime in various states of oxidation, is used for raising up the low lands on the Mersey and forming a foundation for future buildings. The drainage of lands thus treated is offensive: it has a yellow colour, and on exposure to air gives out the gas complained of. At certain spots the streams meet with acid streams, and the gas is then given out in enormous quantities. I have observed one spot, but I believe there must be others underground, perhaps also over-ground.[8]

It was not unknown for the waste deposits to smoulder and burst into flame on a regular basis, producing copious quantities of sulphur dioxide that added to the cocktail of toxic pollutants in the air.

The next major process for extracting sulphur from the waste was developed by Ludwig Mond and patented in 1861. Mond was born in Cassel, Germany and studied with Hermann Kolbe in Marburg and Robert Bunsen in Heidelberg. Like Gossage, Mond had an inventive and tenacious mind that he readily applied to solving technical problems. Mond's process involved blowing air through the sulphur waste to convert the calcium sulphide into hydrosulphide, and then precipitating sulphur by treatment with excess hydrochloric acid (available in large quantities from acid towers).[9] Although Mond claimed that the process could recover 50–60 per cent of the sulphur, in practice the amount was closer to 40 per cent.[10] Mond hoped that, with the support of Angus Smith, the process could 'be adopted by law to prevent the loss of sulphur ... just as the 1863 Act had tackled the release of hydrogen chloride gas with the adoption of the Gossage (or acid) tower'.[11] Unfortunately, it was expensive to operate and was not sufficiently reliable to be incorporated formally into the regulations of the Alkali Inspectorate, although in 1869 the Rivers Pollution Commission did approve Mond's process.[12]

---

[8]   Robert Angus Smith, *Intermediate Report of the Chief Inspector, 1863 and 1874, of his proceedings since the passage of the latter Act*, P.P. 1876 (165), xvi, p. i.

[9]   Haber, *Chemical Industry*, p. 99.

[10]   Ibid.

[11]   Peter Reed, 'Entry for Ludwig Mond', *The Dictionary of 19th Century British Scientists* (Bristol, 2004), pp. 1416–19.

[12]   Ibid.

Another approach to regenerating the sulphur from waste was developed in 1871 by James McTear, manager at Tennant's St Rollox works, near Glasgow. It involved pumping the liquors from the waste heaps into special vessels, treating them with sulphurous acid, and then precipitating the sulphur with hydrochloric acid. It proved reliable in operation and cheap to install; it was widely adopted by manufacturers, even though it recovered only between 27 and 30 per cent of the sulphur. Meanwhile, the sulphur waste was dumped in increasing quantities. According to the 1885 annual report of the alkali inspector, 'There were nearly four and a half million tons of alkali waste in Lancashire alone … it was increasing at the rate of 1,000 tons a day.'[13]

The best process did not emerge until 1888, when the Claus–Chance process was developed; in part, it was based on Gossage's process of 1837. In the Claus–Chance process, ferric chloride catalysed the liberation of sulphur from sulphuretted hydrogen in a Claus kiln, invented in 1882.[14] It was found that 60–80 per cent of the sulphur in the waste could be recovered, making it a much more effective process than its predecessors.[15]

Sulphur waste remained very resilient in the environment from the nineteenth century and its presence survives to the present day even though it is now in a chemically stable state. Although the Muspratt works on the banks of the St Helens Canal at Newton-le-Willows was dismantled in 1851, the mounds of sulphur waste were still present in the early 1980s and were cleared only as a result of land reclamation schemes.[16]

The Dry Copper Industry

In many of his annual reports to Parliament, and using other opportunities as they arose, Angus Smith drew attention to pollution from the dry copper industry, in which copper-bearing ores were smelted. In Britain, the two main centres were Swansea and St Helens, but it was the scale of copper smelting in the former area that highlighted the pernicious and damaging pollution from 'copper smoke'. Angus Smith felt that the Alkali Inspectorate should be more engaged in exploring a technical remedy to reduce the impact of this pollutant.

Copper smoke had a variable composition. Cornish copper ore produced a most exotic mix of environmentally damaging ingredients. It comprised two main components – an 'acid rain' and a fine particulate part. As the Cornish

---

[13]   *21st Annual Report of the Alkali Inspector*, P.P. 1885 (C. 4461), pp. 10–11.

[14]   Haber, *Chemical Industry*, p. 100.

[15]   Alexander M. Chance, 'The Recovery of Sulphur from Alkali Waste by Means of Lime Kiln Gases', *Journal of the Society of Chemical Industry*, 50 (1931), pp. 151–61.

[16]   Even today, large quantities of waste can still be found in Widnes, usually just below the surface of golf courses or any open ground, while the A562 main road between Liverpool and Widnes is very undulating in places due to the underlying sulphur waste.

ores had a high sulphur content, and contained fluorspar, the resulting acid rain contained sulphurous, sulphuric and hydrofluoric acids.[17]

The smoke has been described as the 'acid rain scandal of its day'.[18] The particulate part of the smoke contained copper, sulphur, arsenic, lead, antimony and silver, and was deposited over the countryside within a several-mile radius of the copper works. It was this component that was inhaled by the furnace workers and others working in the smelters, as well as by the local population living within the area of release.

During the nineteenth century, the local population of Swansea had a rather fickle relationship with the industry and the smelter owners. From the beginning of smelting in 1720, there were certainly court cases in attempts to halt or curtail operations, or to seek financial redress for damage caused by the copper smoke. However, legal challenges in the nineteenth century were relatively few. The main reason was that the local population accepted the inconvenience of the copper smoke because the industry provided employment in the Swansea Valley.[19] This was quite unlike the situation with soda works, where the affected landowners and courts of law had dismissed the wider economic interest of the country as a legitimate argument.[20] Nevertheless, some technical innovations were introduced, in part to help avoid the threat of litigation, and in part to enable recovery of useful by-products. However, success came almost too late.

Like soda works, the smelters from the early period used tall chimneys to disperse the smoke more widely, but an additional advantage was that the long flues connecting the furnaces to the chimneys provided cooling surfaces on which some of the sulphur, arsenic, antimony and hydrofluoric acid could condense.[21] Vivian and Sons, one of the largest smelters, had realized by the 1820s that the copper-smoke problem might be reduced by redesigning the furnaces. Most of the copper smoke was produced during the initial calcining stage,

---

[17]   Edmund Newell, 'Atmospheric Pollution and the British Copper Industry, 1690–1920', *Technology and Culture*, 38 (1997), p. 660. The mining industry of Saxony had to tackle the same problems as experienced in South Wales, for which see Franz-Josef Brüggemeier, *Das unendliche Meer der Lüfte: Luftverschmutzung, Industrialisierung und Risikodebatten im 19. Jahrhundert* (Essen, 1996).

[18]   Ralph A. Griffiths, *Singleton Abbey and the Vivians of Swansea* (Llandysul, 1988), p. 32.

[19]   Tal Golan, *Laws of Men and Laws of Nature* (Cambridge, MA, 2004), pp. 70–76.

[20]   Newell, *Atmospheric Pollution*, p. 668. For a history of the copper industry in the Swansea area, see Edmund Newell, '"Copperopolis": The Rise and Fall of the Copper Industry in the Swansea District, 1826–1821', *Business History*, 32/3 (1990), pp. 75–97.

[21]   Antimony was used with silver for yellow and orange glazes in Staffordshire ware, following Wedgwood's investigations, in mirror manufacture, and in alloys such as type-metal, Britannia-metal and plate-pewter. Arsenic was used in glass manufacture, in printing, and as a poison. Often, these substances ended up as waste deposits on land surrounding the copper works.

when large quantities of sulphur were removed. Although the chemistry was straightforward, designing a suitable furnace proved much more challenging – the coal smoke prevented the oxidation necessary to produce sulphuric acid. At various times, both Humphry Davy and Michael Faraday were consulted on the design and operation of suitable furnaces.

Two types of furnace were developed in an attempt to reduce copper smoke, and particularly its sulphur component:

> The first to be developed was the muffle furnace, which kept the copper smoke fumes separate from the coal smoke, by heating the ore, which was laid out on a flat bed, from underneath the bed, there being no direct contact between the ore and the heat source ... The second type of furnace, developed by Moritz Gersteinhöfer in Germany, where smelter smoke was also becoming a serious problem, produced very little coal smoke, as it required fuel only to begin calcination. Once ignited, the furnace used the sulfur in the ore as fuel, thus eliminating the presence of coal smoke and indeed significantly reducing the amount of coal required in calcination.[22]

Although the upbeat announcement of these furnaces was warmly welcomed in the expectation of some marked reduction in copper smoke, it was premature because in operation the furnaces proved to have limitations.[23] From 1886 attempts were made to use an electrical precipitation technique in the flues, although this development was not used extensively until after the First World War, by which time copper smelting in Britain was in its terminal phase.

Angus Smith, with his interest in acid rain, was particularly concerned about the very damaging effect of smoke from the copper smelters.[24] He felt that more could have been achieved in operating the redesigned furnaces and in making better use of the flues for condensing out smoke-borne materials. However, there was a major obstacle: whereas the soda manufacturers had reluctantly agreed to the inspection regime, Hussey Vivian, the influential owner of the major copper smelter in Swansea, was totally opposed to any inspector entering his works and offering recommendations. This left no room for compromise, and the 1876 Royal Commission on Noxious Vapours backed off from making a recommendation to include inspection of copper smelters; Parliament followed suit in its 1881 legislation.[25]

---

[22]   Newell, *Atmospheric Pollution*, p. 677.

[23]   Ibid., p. 678.

[24]   Smith was the first to use the term 'acid rain' in 1859, in his paper 'On the Air of Towns', *Quarterly Journal of the Chemical Society*, 11 (1859), p. 232.

[25]   See Hussey in evidence to the 1876 Royal Commission on Noxious Vapours. *Royal Commission on Noxious Vapours, II. Minutes of Evidence*, P.P. XLIV (1878), pp. 450 and 452.

Cement Works

During the course of their work, the inspectors' attention was often drawn to pollution from cement works, particularly those for Portland cement.[26] Portland cement, one of the most important types of cement because of its likeness to Portland stone quarried on the Isle of Purbeck, was made by heating together lime and clay in a specially constructed kiln at a temperature of about 1300 °C. There were several problems as far as atmospheric pollution was concerned; these included dust and gas emissions (there were also concerns about fuel economy, because the dust and gases were emitted at high temperature). In his evidence to the Royal Commission in 1876, William Odling, of the University of Oxford, who had considerable experience and knowledge of the manufacture of Portland cement, expressed his view that

> At the present time all emanations which are given off from the kilns are given off from the top of the chimney shafts, and these emanations consist, first of all, of cases, which, of course, are invisible, and then of products which are visible, or which are recognizable by the smell.[27]

The dispersion of the offending pollutants followed the approach used with hydrogen chloride gas, namely the very tall chimneys, which had proved just as ineffective. Offending smells also originated from the organic matter in the clay.[28] In addition, a discernible white vapour escaped from the works that included fine cement dust (from mechanical processing) and volatile salts such as potassium chloride, sodium sulphate and potassium sulphate.

Odling thought that the smell of the burning organic matter could be removed by complete combustion in the kiln, and the other pollutants at least reduced by use of the Hoffmann kiln, which had yet to be tried on a large scale.[29] Angus Smith was also of the view that these kilns would play a part in reducing the pollution. He made the important point that, although cement works should not be included in the list of works for routine regulation, where a complaint was made the inspectors should have the power to include them to ensure that the processes were being operated in the most effective way to reduce pollution. This approach was to be one of Angus Smith's key proposals for improving the future working of the legislation.

---

[26]   A.R. Meetham, *Atmospheric Pollution* (Oxford, 1956), pp. 92–3.

[27]   *Royal Commission on Noxious Vapours*, II, p. 392.

[28]   Ibid.

[29]   Many of the same problems were reported in W.A. Damon, 'The Alkali Act and the Work of the Alkali Inspectors', *Royal Society of Health Journal* (London), 76/9 (1956), pp. 566–75. An early Hoffmann kiln is preserved at Betchworth, Surrey.

Potteries

Potteries were invariably associated with high levels of pollution. One inhabitant in the neighbourhood of the pottery towns felt that, on foggy days, there was little to choose between the atmosphere of Hanley or Longton, in Staffordshire, with that of Widnes.[30] Much of the pollution was due to coal-burning for heating the kilns: where there was a concentration of kilns, the atmosphere proved very difficult for people living close to the potteries. Several potteries were referred to the Alkali Inspectorate, not primarily because of the smoke, which lay outside their jurisdiction, but because of other obnoxious vapours emanating from the kilns. Targets of complaint were kilns producing salt-glazed pottery. In the final 20 to 40 minutes of firing the kiln, there was a release of hydrogen chloride gas after common salt was thrown over the pottery in its red-hot state to produce the glaze. As Angus Smith reported in response to a specific question during hearings of the 1876 Royal Commission on Noxious Vapours, 'It probably occurs, twice, and not, I think, so much as thrice a week, for every kiln, and where there are many kilns, it may occur frequently in the week.'[31]

While the atmosphere was a nuisance to local people, the pottery workers were also exposed to other dangers, ' inhaling dust from siliceous and common clay, flint, Plaster of Paris, calcinated bones, and feldspar', in addition to the harmful fumes.[32] Mortality rates, first studied during the second half of the nineteenth century, showed that male pottery and earthenware workers had one of the highest mortality rates of any occupational group; it was almost twice the rate for coal miners in South Wales (42.97 per thousand, as compared with 24.27 per thousand).[33] Although not the responsibility of the Alkali Inspectorate, occupational diseases and public health concerns were to become increasingly important factors in the operation and control of chemical processes. It was only quite late in the century that these health components were brought together in an administrative and regulatory framework.

Ammonia and Fertilizer Works

The development of the coal-gas industry produced large quantities of ammoniacal liquor, which was usually run to waste, at least until the 1870s, when it was converted into ammonium sulphate and used as a fertilizer.[34] By 1886 some

[30]   A.E. Fletcher, 'Modern Legislation in Restraint of Emission of Noxious Gases from Manufacturing Operations', *Journal of the Society of Chemical Industry*, 11 (1892), p. 124.

[31]   *Royal Commission on Noxious Vapours*, II, p. 6.

[32]   Anthony S. Wohl, *Endangered Lives: Public Health in Victorian Britain* (London, 1974), p. 274.

[33]   Ibid., pp. 280 and 281.

[34]   F. Sherwood Taylor, *A History of Industrial Chemistry* (London, 1957), pp. 209–10.

82,500 tons of ammonium sulphate were made from the liquor and production rose to 142,400 tons by 1900 to become a major part of the fertilizer industry.[35]

Reports of releases of ammonia from ammonium sulphate works were regularly reported to the Alkali Inspectorate, and were used by Angus Smith in evidence before the 1876 Royal Commission on Noxious Vapours.[36] In most cases, the ammoniacal liquor was pumped from the distant gasworks to the ammonium sulphate works; sometimes the liquor could not be treated at the same rate as it was pumped from the gasworks.[37] As a result of occasional incidents, the Royal Commission on Noxious Vapours recommended that ammonium sulphate works should be placed 'under inspection', and that there should be 'adoption of the best possible means'. When such a means was developed, the Alkali Inspectorate could ensure that it was utilized at all ammonium sulphate works.

Chemical Manure Works

Liebig had proposed the use of artificial manures following his work analysing the minerals that plants took from the soil, but the British patent that James Muspratt took out on behalf of Liebig in 1845 failed to produce a workable fertilizer because the minerals incorporated in the artificial manure did not readily leach into the soil.[38] It was Liebig's ideas as presented in his later publication, *An Address to the Agriculturalists of Great Britain*, that had prompted John Lawes's experiments at his Rothamsted family home and led to Lawes's patent in 1842 for the more soluble calcium superphosphate fertilizers, or chemical manures, as they became known.[39] The essential part of the manufacture was to dissolve the phosphoric substance in sulphuric acid to create a paste-like material that was then absorbed in an inert carrier material.[40] These absorbent materials came in many different forms, including

> bran sawdust, dust of malt, husks of seeds, brewer's and distillers' grain, ground
> rags, pulverised rope or linseed cakes, the refuse of flax leaves, bark, dry tan,

---

[35]   Haber, *The Chemical Industry*, pp. 105–6.

[36]   *Royal Commission on Noxious Vapours, I. Report*, P.P. XLIV (1878), p. 25.

[37]   In one incident a person died when a blocked sewer became contaminated by the flow of ammoniacal liquor.

[38]   The collaboration with James Muspratt came about from their meeting during the 1837 British Association for the Advancement of Science meeting in Liverpool. The intention was to trial the manufacture in Liverpool and then expend into Germany and other countries. The patent was BP 10,616/45.

[39]   Justus von Liebig, *An Address to the Agriculturalists of Great Britain* (Liverpool, 1845). Lawes's patent was BP 9353/42. Lawes developed part of the family estate at Rothamsted into an agricultural experimental station.

[40]   Alex Campbell, *The Chemical Industry*,(London, 1971), pp. 75–80.

siliceous sands, peat, or other sandy mould, dry dust, earth or clay, fine sifted cinders, ashes and the like.[41]

When manufacture got under way a variety of phosphoric substances was used, including bone, bone ash and coprolites; later coprolites were imported from Spain, Canada and South Carolina (USA).[42] The works were often accused of creating a nuisance because of the appalling smells emanating from their operations, not so much in manufacturing the superphosphates but more in the processing of the animal materials to generate the bones. While these chemical manure works were scattered around the country, several of the large-scale manufactories were located on the banks of the Thames estuary where the smells affected people living in Greenwich and Blackheath. During the hearings of the 1876 Royal Commission on Noxious Vapours, evidence was taken from both the manufacturers about their operations and from the local residents and officials about their attempts to have the vapours controlled.

While the Lawes Patent Manure Company had a works in the vicinity, the main attention was focused on two companies, Gibbs and Company and Odams and Company, both operating on the north bank of the Thames and on the bank opposite both Greenwich and Blackheath. In his evidence, James Odams explained how the nature of the processes had changed. When the works began operations in 1851, 'from three to four thousand gallons of blood per day in its raw state' were used but the use of blood was discontinued. Bone boiling, with its obnoxious extraction of various fats, had also taken place earlier but not from the 1870s, and Sir John Morris, a chemical manure manufacturer in Wolverhampton and Chairman of the Chemical Manure Manufacturers' Association, agreed that in an earlier phase materials such as 'dead animals, putrid carcases, rotten eggs and any waste animal matter' formed part of the production.[43] The main operations involved crushing the bones, treating the bone dust with sulphuric acid and then absorbing the superphosphates into some inert materials. The process comprised

Firstly, a Schiel's fan of large size, driven by steam power. Secondly, a trough or conduit, for leading the gases from the mixers and manure chambers; this trough is called the main trough. Thirdly, two small troughs or leads, one from each mixer to the main trough. Fourthly, a pair of leaden receivers or condensers, with water supply. Fifthly, two small troughs or leads from the condensers (one from each) to the smoke shaft.[44]

---

[41]   W.A.L. Alford, and J.W. Parkes, 'Sir James Murray: a pioneer in the making of superphosphate', *Chemistry and Industry*, 1953, p. 852.

[42]   Coprolites are fossilized dung; British sources were found in Cambridgeshire and Suffolk.

[43]   *Royal Commission on Noxious Vapours*, II, pp. 345 and 347.

[44]   Evidence of James Odams, *Royal Commission on Noxious Vapours*, II, p. 347.

Attempts were made to retain any vapours within the mixer:

> There are two mixers used alternately. They are of a similar construction. The mixer is entirely enclosed with the exception of the aperture at the side for introducing the materials to be operated upon. This aperture has a canvas curtain, renewed from time to time, and there is found to be no escape of gas from this opening, the suction caused by the fan instantly drawing into the small mixer trough, and thence into the main trough, any gases evolved, while any 'puffing out' of the gases at the aperture is prevented by the curtain before alluded to.[45]

The total production of chemical manures in Britain at the time of the Royal Commission was about 600,000 tons per annum.

Those affected by the smells in Blackheath and Greenwich spoke of how they had suffered headaches, sickness and loss of appetite. John Spencer, a Blackheath solicitor, gave detailed evidence about action taken through the Local Boards to try to have the chemical manure works controlled or even closed down through the 1855 Nuisances Removal Act, but magistrates were very reluctant to act and legal action had made no difference.[46] Often in making any accusation it was necessary to state quite categorically which works was causing the nuisance, but this had proved very difficult given the different manufactories in close proximity and the mixture of smells, which varied over time. Since the existing sanitary regulations did not allow inspectors to enter works, many of those affected felt that inspection by the Alkali Inspectorate was needed, although the Commissioners raised concerns about which gases were to be regulated and what amounts.[47] In their report the Commissioners said that chemical manure works should be required to adopt the 'best practicable means' in preventing the escape of their obnoxious vapours and that the works should be subject to inspection.

Sulphur and Nitrogen Acid Gases and Coal Smoke from Salt Works

From the time that he was appointed head of the Alkali Inspectorate in 1864, Angus Smith was well aware of the release of sulphur and nitrogen acid gases from alkali works. Standard limits for gases other than hydrogen chloride were recommended by the Royal Commission on Noxious Vapours in 1878; they became part of the 1881 Act. The limits set were one grain of sulphur per cubic foot, and half a grain of nitrogen per cubic foot, alongside the limit of 0.2 grains for hydrogen chloride.

---

[45]   Ibid.

[46]   The ineffective action of magistrates was frequently expressed. It should be remembered that magistrates were often manufacturers and were being called upon to act against other manufacturers.

[47]   Ibid.

The damaging effects of sulphur acid gases from the burning of coal had been studied by Angus Smith from the late 1850s. The results of these experimental investigations were included in his 1872 book, *Air and Rain: The Beginnings of a Chemical Climatology*, in which he hoped to take the study of air quality well beyond the limited domain of the Alkali Inspectorate.[48] The pernicious nuisance of coal smoke that blighted the urban environment from the introduction of coal burning in the sixteenth century and prevented people from opening windows to adequately ventilate their living accommodation was always a target for effective control. Even though control remained outside the remit of the Alkali Inspectorate, Angus Smith always looked for an opportunity to engage in the debate over the control of coal smoke. One such opportunity arose in the 1880s, when salt works came under his inspection regime. Although Angus Smith initially hoped that control of smoke from salt works might act as a springboard for the wider control of coal smoke, he backed down when drafting the Provisional Order:

> I was hoping that the best practical means of burning coal at salt works could be introduced. It would be of immense benefit to the country. The number of ways of doing it is very great and it only requires a certain authoritative beginning in salt works and from then it would be spread.[49]

Final resolution of the coal-smoke problem would have to wait another 72 years before effective government legislation was put in place.

Protests by Campaign Groups and the Public

Parliament's approval of the 1863 Alkali Act and the subsequent setting up and working of the Alkali Inspectorate did not end the public protests about the damaging effects of hydrogen chloride gas, which was continuing to be released. It seemed to many that the escapes were as large as in the period before the legislation, and even though Angus Smith was able to show in his annual reports condensation levels around 98–9 per cent, the 1 or 2 per cent of escaping gas still represented a sufficiently large quantity that could damage the natural environment and create unhealthy living conditions.

For its part, the Alkali Inspectorate was limited by the number of inspectors with which to cover an increasing number of alkali works across the country, and there was little likelihood of government approval for additional inspectors at least in the short term. Visits by inspectors were based in part on previously reported escape episodes and these allowed close monitoring of the operation of plant to ensure that fundamental flaws did not occur. Many of the larger escapes occurred at night, when inspectors were reluctant to visit works and scramble around plant in darkness or low light levels.

---

[48] Smith, *Air and Rain*, pp. 464–8.

[49] Letter from Robert Angus Smith to Local Government Board, 21 January 1884. NA (Ref: MH. 16/2).

Automated aspirators were introduced to collect gas samples over many days, allowing the inspector time to visit other works. There is some evidence that escapes often took place at weekends in pursuit of increased production by piece-meal working – again, when inspectors were unlikely to visit. Under Angus Smith's influence responsibility was placed firmly back on the manufacturers through the works manager and the foreman.

Where there were persistent escapes, individual members of the public came together to form campaign groups; many of these groups grew out of campaigns to reduce pollution from coal smoke since the distinction between the two forms of pollution was too subtle when both harmed the natural environment and blighted people's lives. Probably the first of these groups was the Lancashire and Cheshire Association for Controlling the Escape of Noxious Vapours and Fluids from Manufactories, established in Warrington in 1874.[50] In due course several allied branches of the Association were set up, including Liverpool, Northumberland and Manchester.[51] The Manchester and Salford Noxious Vapours Abatement Association was formed at a meeting in Manchester Town Hall on 2 November 1876 and its role was 'to procure improved legislation for the control of Alkali and other Chemical Works, and to disseminate information as to their injurious influence on health and vegetation'.[52] The Manchester Association was not just a campaigning group but sought to investigate the effects of chemical vapours on health and take direct action by monitoring alkali works and noting any escapes. At a meeting on 1 March 1877 a Salford surgeon was appointed as inspector at an honorarium of 20 guineas, but this initiative proved short lived because the inspector was not felt to be carrying out his duties thoroughly enough. Interestingly, unlike many other similar associations, the Manchester Association originated with fighting chemical vapours and then on 1 March 1877 resolved to include suppression of smoke nuisance in its objects.[53] Such an extension to its activities led it to rent accommodation in which to display 'smoke consuming apparatus' and meet with the mayors of Manchester and Salford to seek 'energetic action on the part of the Corporation of each borough in dealing with smoke and chemical vapour nuisance'. Realizing that insufficient was being done to monitor the situation, the Association resolved to approach the Manchester and Salford Sanitary Association for financial support towards appointing an inspector, but they were unable to assist.

When the 1876 Royal Commission on Noxious Vapours was taking evidence, landowners and their land agents and various campaign groups took every opportunity to inform the commissioners of their experiences. The list of landowners and land agents giving evidence reads like a who's who of the wealthy in areas affected by

---

[50]   *Minute Book of Manchester and Salford Noxious Vapours Abatement Association*, GMCRO (Ref: M126/6/1/1).

[51]   Wilmot, 'Pollution and Public Concern', p. 124.

[52]   *Minute Book of Manchester and Salford Noxious Vapours Abatement Association*, GMCRO (Ref: M126/6/1/1).

[53]   Minutes of Meeting on 1 March 1877, *Minute Book of Manchester and Salford Noxious Vapours Abatement Association*, GMCRO (Ref: M126/6/1/1).

pollution, and included Sir Richard Brooke (Mersey), Sir Armine Morris (Swansea), George Spain (land agent for Sir Webster James in Jarrow), Henry Wallace (land agent to Lord Ravensworth in Tyneside) and Adolphus Benedict Moubert (land agent to Lord Gerard in Newton). When the Lancashire and Cheshire Association for Controlling the Escape of Noxious Vapours and Fluids from Manufactories appeared before a group of commissioners that included Professor Henry Roscoe and Professor Frederick Abel on 19 August, the evidence was lengthy and detailed. First, the Association chairman, William Beamont, put the occurrence of pollution into a historical perspective going back to 1850 when James Muspratt's works at Newton was accused of causing a nuisance, but the core of his evidence was nine general and overarching statements that were then followed up by more detailed evidence given by parties such as Sir Richard Brooke and a number of land agents.

One witness, John Pemberton, who was in charge of one of the horses guiding the flat, *Liberator*, gave evidence of a dramatic incident when the boat was proceeding past the alkali works at Weston (in Cheshire). It is one of the starkest reports of an episode showing the damaging effects of the acid gas on a person. The evidence proceeds:

> Were you in charge of one of the horses towing the Liberator on that morning? – I was.
>
> Did you pass by the alkali works at Weston? – Yes.
>
> You were then upon the canal bank? – Yes.
>
> Were the alkali works close to the towing path? – Yes.
>
> Will you state what happened on that occasion? – I was passing, and I met a man at the lower end of the works.
>
> What happened to you? – I tumbled down from the gas coming from the works.
>
> Did you feel it coming upon you? – Yes.
>
> Did you smell it? – Yes.
>
> How soon did you fall down? – When I had gone about 10 yards into it.
>
> Were you on foot? – Yes.
>
> Did you fall down senseless? – I did.
>
> When you recovered, where were you? – At Sutton Lock.
>
> Was anyone with you? – Yes.

Where were you? – I was aboard the flat, the 'Liberator'.

How long had you been insensible? – Up to Sutton Lock.

For what length of time? – About three quarters of an hour.

How did you feel afterwards? – I felt very bad and sickly. They gave me some stuff.

How long were you in that state? – All the way up the river; for two or three days after.[54]

While most landowners and land agents complained about the damage wreaked on their property, several others gave evidence of little or no damage; this accusation and counter-accusation was one of the main reasons why manufacturers were able to avoid their full responsibility for the extensive damage done until the late 1850s when the acid tower began to be widely adopted. It is important to remember that damage to forests and woodlands might also be caused by poor management of both the land and the trees. Such evidence was often only used during major court cases, involving claims for substantial damages that warranted evidence from a botanist or forestry expert. Such a case was a claim brought against James Muspratt by Sir John Gerard of New Hall, near Garswood, Lancashire in September 1846. William Daniels, gardener to Sir John (and to his father), gave evidence of the damage to the trees in these plantations that had become noticeable in 1837 and 1838, and had deteriorated further during 1839–40 to become a major concern. It appeared that different species of trees were affected to different degrees; alders and sycamores survived well, while oaks were badly affected. Daniels had observed that trees facing the chimney of Muspratt's works at Newton-le-Willows were more badly damaged. Since 1840, he had cut down many thousands of trees of varying age from 5 to 100 years. Other witnesses for the plaintiff gave similar evidence of damage. John Lindley, Professor of Botany at University College London, followed and spoke of the good soil in the plantations and had observed that 'the injury at its outset was most apparent at the extremities of the upper branches'.[55] He was adamant that the damage was linked to the acid gas from the alkali works. In cross-examination, Muspratt's lawyer, Mr Sergeant Murphy, suggested that the 'staghead damage was caused by defects at the roots and that some trees were more susceptible to the acid gas'. To mitigate the damage, witnesses were called to testify that the trees' poor appearance and condition were due to 'great age, exposure to the weather, or to wet soil and want of drainage'. After the jury had deliberated for little more than an hour, they returned with a verdict for the plaintiff and awarded damages

---

54   *Royal Commission on Noxious Vapours*, I, p. 32.

55   'Report of Gerard, Bart. v Muspratt', *The Times*, 3 September 1846.

of £1,000. Muspratt had probably expected the verdict to go against him but was no doubt relieved to find the damages reduced from the original claim of £30,000.

Extension of Regulation to 1881

Royal Commission on Noxious Vapours 1876

During hearings of the Royal Commission, evidence was presented on several individual gases and processes. More important, however, were the several changes to the working of the regulation, all discussed at length. Recommendations in the final report laid down the basis for the subsequent adoption of key elements within the legislative framework on pollution regulation that extended well into the twentieth century. These included 'best practicable means', extension of provision, central administration and enforcement by legal prosecution.

Best Practicable Means

For some time, and certainly from 1876 when giving evidence to the Select Committee on Noxious Vapours, Robert Angus Smith and his inspectors developed an understanding with manufacturers regarding the adoption of the 'best practicable means' for preventing the release of pollutants into the atmosphere. In the case of the Leblanc process, the effective use of the acid tower and other plant was known to allow manufacturers to remain within the limits of discharge for hydrogen chloride gas. Unfortunately, for many other processes, the plant for keeping discharge within the desired limits was not available. To avoid restricting discharge, and waiting for the ultimate plant, it was felt necessary for manufacturers to use the 'best practicable means' then available. As innovations were introduced and more effective plant installed, so eventually the low discharge limits could be achieved. Such an approach would also enable the inspectors to continue their peripatetic consultancy role and assist manufacturers with improving the working of their existing plant, as well as ensuring that more effective plant was developed over time.

This approach was promoted by Angus Smith and his colleagues on a number of occasions during their evidence to the 1876 Royal Commission on Noxious Vapours, and the Commission's Report drew attention to the necessity for recording whether the 'best practical means' was applied:

> They [the sub-inspectors] should make monthly reports of the regular inspection of the works in their districts, according to forms prescribed by the Local Government Board; and their reports should include statements as to whether the 'best practicable means' are employed at works where limits of acid escape are not prescribed, as well as the results of such examinations of vapour escapes

as are made from time to time at such works. They should also report any case of important accidental failure to comply with the prescribed test, and of neglect to apply the 'best practicable means' to the prevention of vapour escapes.[56]

This approach was felt to be the most effective way of reducing harmful pollutants over the long term while allowing the Alkali Inspectorate to draw up a list of pollutants and processes, and stipulate where rights of entry for inspection were necessary. However, Angus Smith was adamant about the ultimate use of effective plant: 'My opinion is almost exactly as it was some years ago that the moment that we can obtain the proper apparatus the law should be carried out with its upmost rigidity.'[57]

Extension of Provision

Inevitably with the rapid expansion of the chemical industry during the second half of the nineteenth century the Alkali Inspectorate became increasingly concerned about the time lag between notification of a pollutant or polluting process and incorporation of such notification into the regulatory legislation. Powers had to be more readily available to the Alkali Inspectorate, the Local Government Board or the minister to reduce the time lag to an acceptable level. This requirement was recognized by the commissioners in 1878, and provision was made in the 1881 Act for the minister to make Orders 'extending the list of defined scheduled processes and the list of noxious and offensive gases'.[58] This ensured that the Inspectorate was seen more often than not taking the lead rather than lagging behind and facing numerous protests.

Central Regulation – For and Against

With the absorption of the Alkali Inspectorate into the Local Government Board following the 1872 Public Health Act, there were regular discussions and disagreements within the Board on whether the function of the Inspectorate should continue to be centrally organized or whether it should be carried out at the local level as a function of local authorities. In reviewing some of the options in an attempt to find a compromise, there was even the suggestion of forming groups of districts serving areas where the industry was prevalent, who would appoint and manage the sub-inspectors but with agreement and oversight from the Chief Inspector, to whom reports would be sent on a regular basis. This option was proposed in the context of machinery agreed as part of the 1875 Public Health Act for the 'union of districts for certain purposes'.[59] Angus

---

[56]   *Royal Commission on Noxious Vapours*, I, p. 28.
[57]   *Royal Commission on Noxious Vapours*, II, p. 479.
[58]   Damon, *Alkali Act*, p. 567.
[59]   *Royal Commission on Noxious Vapours*, I, p. 30.

Smith remained consistent and adamant about the advantages of central control: consistency of regulation, well-qualified staff with the appropriate scientific and technical knowledge and experience, cost-effective administration and greater independence from vested interests. The manufacturers, from their positions of limited concern and lukewarm support in the early years of the legislation, lent support to Angus Smith's stance. However, opposition to his position within the Local Government Board was formidable, with George Sclater-Booth (President), John Lambert (Secretary) and John Simon (Medical Officer of Health) all supporting regulation through the local authorities. While giving evidence to the 1876 Royal Commission on Noxious Vapours, John Simon was very clear about his position, supporting inspection by local authorities, with central government being responsible for inspectors of nuisance authorities.[60] This is hardly surprising, given the overriding role of the Board in delegating responsibilities for government activities, as Roy Macleod has observed:

> It was obvious that a system of local government inspection had the advantage of being in harmony with recent legislation and the extension of local responsibility and power, but it was clear that many local authorities did little to enforce laws which would adversely affect the interests of their most important constituents.[61]

There was the added difficulty of impartiality on the part of local authorities since many of the councillors were involved in commerce or industry, and therefore had a vested interest.

In their final report, the commissioners supported Angus Smith and central administration, much to the disappointment and chagrin of the Local Government Board and its principal officers. Later, when additional inspectors were requested by some local authorities because of the scale of pollution, the cost was borne wholly or in major part by the local authority, with inspectors based locally but always deemed part of the Alkali Inspectorate and reporting to the Chief Inspector when enforcing the regulations and undertaking their inspection duties.

Enforcement by Legal Prosecution

From his initial appointment as Inspector in 1864 Angus Smith had taken the view that enforcement of the terms of the 1863 Alkali Act would be more widely achieved by working in a cooperative way with the manufacturers, with the team of sub-inspectors becoming peripatetic consultants. This decision was made in all probability because of the manufacturers' lukewarm support for inspection during their evidence to the Select Committee on Injury from Noxious Vapours in 1862. But another important contributory factor was Angus Smith's experience as an expert witness for the defence in the *Regina v. Spence* court case in 1857 that led

---

60   Ibid., p. 26.
61   Macleod, 'Alkali Acts Administration', p. 103.

him to question the effectiveness of legal prosecution as a means of regulating industrial processes.[62]

When the commissioners heard evidence during the 1876 Royal Commission on Noxious Vapours they were constantly told by witnesses of ineffectual enforcement of the regulations on the part of the Inspectorate, claimed to be causing further environmental damage. When Angus Smith was called to give evidence, he was asked how many prosecutions had been made and the answer was four. This shocked the commissioners and their final report made very clear that with the large number of gas escapes much greater use should be made of county court prosecutions to bring redress. With the routine accusations of regulatory failure made to the Local Government Board and concern expressed to the Royal Commission, John Lambert (Secretary of the Board) became impatient with Angus Smith, and instructed him to undertake prosecutions as a means of enforcement. Angus Smith remained very reluctant to proceed in this way, feeling that such legal action would undermine the good working relationship he had created with the manufacturers. Accordingly, whenever a situation with a manufacturer looked as if it might lead to prosecution he always ensured that the Local Government Board took the final decision. One such case occurred soon after publication of the Royal Commission's report involving Snape and Company of Widnes who had accidently allowed an escape of the acid gas. With pressure from the Local Government Board it was agreed to start proceedings. Snape and Company responded with a profuse apology and requested that the prosecution be stopped, but Lambert felt the need to show iron resolve:

> The Board regrets the necessity of proceeding in these cases but so much blame has been incurred by the supposed indisposition of the Board to enforce the provisions of the Act that they do not feel justified in dispensing with these actions.[63]

Given such determination on the part of the Local Government Board, Angus Smith was forced to continue with the prosecution. Without waiting further, Snape and Company responded by paying the maximum fine of £50, and as a consequence the case was dropped. This pattern of actions proved to be a regular occurrence in such cases, so while more prosecutions were threatened, most companies paid the fine and avoided having their name and reputation tarnished in court. Nevertheless, by involving the Local Government Board in the decision to prosecute, Angus Smith was able to retain for the most part his good working relationship with the manufacturers.

---

[62]   Golan, *Laws of Men*, pp. 76–80.
[63]   Letter from John Lambert to Robert Angus Smith, dated 22 October 1878. NA (Ref: MH 16/1).

Link with Public Health and Occupational Health

The early court cases seeking damages for pollution from alkali works by hydrogen chloride gas were focused on effects on the environment rather than on chemical workers and local inhabitants. Although evidence was taken from medical practitioners, the information presented concerned little more than discussion of the concentrations of hydrogen chloride in the air at which people would find their breathing impaired or might be asphyxiated. Such evidence by one doctor was often challenged by another, and as a result the medical evidence remained contentious, and was often excluded by judges and juries when reaching their verdicts. Although it might be assumed by all parties that air contaminated with hydrogen chloride would have a detrimental effect on health, there were many people who felt that such air had disinfectant properties. During a court case in Liverpool in 1831 concerning the alkali manufacturer James Muspratt, supporters had erected a banner on the chimney of his works proclaiming such benefits against major contagious diseases such as typhus, and proposed that hydrogen chloride gas be pumped into every home in the town because of its perceived benefits. This was not an isolated example, and such claims continued well into the 1870s. But what was the role of the public health movement and the concerns of occupational health as air quality continued to decline through the nineteenth century? The Public Health Act of 1848 required major towns and cities where the death rate was greater than 23 per 1,000 to set up public health boards and to appoint a medical officer of health.[64] The responsibility of the medical officer of health was to prevent the spread of highly contagious diseases, such as typhus and cholera, among those housed in very cramped, unhygienic conditions.

One of the most important instruments used by medical officers of health was the statistical information collected on deaths from contagious diseases, based on the work of William Farr.[65] The Registration Act of 1836 required the General Register Office to gather statistics on pollution. Farr took charge of the statistical abstract.[66] His concern was the cause of death:

> Diseases are more easily prevented than cured, and the first step to their prevention is the discovery of their exciting causes. The registry will show the agency of their causes by numerical facts, and measure the intensity of their influence.[67]

---

[64]   The first medical officer of health was William Henry Duncan, in Liverpool, appointed in 1847 following a separate parliamentary bill arising from the extreme conditions existing in the city following the spread of typhus.

[65]   William Farr was appointed to the General Register Office in 1839.

[66]   Wohl, *Endangered Lives*, pp. 29–31.

[67]   *First Annual Report of the Registrar-General*, Appendix (P), P.P. 1839 (187), pp. 64 and 65.

Although the main remit for the medical officers of health was to prevent major contagious diseases, once air pollution from chemical processes grew rapidly they began to be consulted more regularly on the public health implications of the pollution. The case of copper smoke in the Swansea area provides a good example of where the views of medical officers of health were sought, but they were found to be very contradictory over time. These initial informal consultations later developed into more robust deliberations that led to an increasingly important role in the regulation of chemical processes alongside the Alkali Inspectorate.

Nevertheless, as in the case of hydrogen chloride gas from the Leblanc works, many in the medical profession felt strongly that copper smoke acted as a prophylactic against contagious diseases such as cholera. Henry Vivian, the Swansea copper smelter, was so convinced of the medical efficacy of the sulphurous copper smoke that he dispensed dilute sulphuric acid to his workers to ward off a cholera epidemic in the Swansea area. This was despite the fact that the distressed visual appearance of workers in the industry was clear for all to see.

Under the direction of William Farr, the collating of medical statistics became more sophisticated, and took account of smaller geographical areas. Even so, the general sanitary conditions in Swansea were so grim that they tended to mask the effects of the copper smoke in the small pockets of villages surrounding the town. It should also be noted that the Swansea Board of Health paid 'little attention to copper smoke', a stance that acknowledges again the tremendous economic importance of copper smelting to this area of South Wales. This stance may also reflect the fact that, without a causal link, the Local Government Board was more preoccupied with major epidemics and the low level of general sanitation.[68] In the early part of the nineteenth century, few if any attempts were made to investigate the causes of industrial diseases, even though they were often threats to life. It was only around 1890 that methodical attempts were made to identify occupational hazards, record their occurrence and associated mortalities, and mitigate their impacts on the health of workers.

The chemical industry provides a good example of the growing awareness of occupational health. By the census of 1911, the industry employed about 171,983 people in England and Wales – the industry was defined as two sectors: the alkali industry and associated processes; and the fine chemicals and drugs industry.[69] The alkali industry provides two stark illustrations of hazardous working conditions. The first regards the furnace-men who were responsible for the salt-cake furnaces, where large quantities of hydrogen chloride were produced. Here, 'the workmen either wear a flannel muffler tied over his face,

---

[68]   Newell, *Atmospheric Pollution*, p . 683.

[69]   E.W. Hope, in collaboration with W. Hanna and C.O. Stallybrass, *Industrial Hygiene and Medicine* (London, 1923), p. 488.

or he bites a piece of flannel between his teeth and breathes through it'.[70] Nevertheless, no matter how thick the flannel, the furnace-men's teeth quickly rotted away. The second hazard was associated with the bleach packer, again using simple flannels, goggles and 'a piece of paper round his trousers to keep the bleach from attacking them'.[71] At least this was the case until 1888, when manufacturers began to adopt the Hasenclever chamber.[72]

As concerns increased and monitoring became more thorough, so the effects of occupational hazards on those working in the chemical industry began to emerge. E.W. Hope's *Industrial Hygiene and Medicine* provided some general observations from the information collected, including about respiratory diseases, 'fatal accidents among which are to be reckoned those deaths due to gassing', and cancer mortality 'considerably above the average'.[73] It is not too surprising that, given the nature of their employment, chemical workers had 'a high mortality from diseases of the liver and other digestive diseases' due to alcohol abuse.[74]

By the 1890s, it had become possible to compare mortalities from specific causes for different occupations and also for different age groups.[75] Further legislation and additional inspectors reinforced the attention given to occupation diseases: the 1895 Factory and Workshop Act stipulated the notification of industrial diseases for the first time; in 1896, Dr Arthur Whitelegge was appointed as Chief Inspector of Factories; and in 1898, Thomas M. Legge was appointed Medical Inspector of Factories.[76]

Alkali, etc., Works Regulation Act 1881

The recommendations in the final *Report of the Royal Commission on Noxious Vapours* published in 1878 were very much in line with the issues Angus Smith had addressed during his evidence to the Commission. Unfortunately, there was a downside – the cordial and supportive relationship between Lambert and Angus Smith that had enhanced the workings of the Alkali Inspectorate was at an end as a result of the criticism levelled at the Local Government Board during the proceedings of the Royal Commission and the divergent views on central administration of the Inspectorate and on the role of prosecution in enforcing regulation. As if in pique, Lambert refused Angus Smith's travel request to ascertain developments in continental regulation and later refused to consider a salary increase for the

---

[70]   A.P. Laurie, 'The Chemical Trades', in Thomas Oliver (ed.), *Dangerous Trades* (London, 1902), p. 571.

[71]   Ibid., p. 576.

[72]   Campbell, *Chemical Industry*, p. 68.

[73]   Hope, *Industrial Hygiene and Medicine*, p. 489.

[74]   Ibid.

[75]   Laurie, 'Chemical Trades', pp. 594 and 595.

[76]   Wohl, *Endangered Lives*, p. 263.

inspectors.[77] However, as Roy MacLeod has pointed out, Lambert's greater resolve towards holding companies to account ensured that he was firmly behind Angus Smith's wider coverage of regulation. But manufacturers were concerned about hasty legislation that might threaten the economic prosperity of the country and on 5 January 1879 a deputation from the Alkali Manufacturers' Association met with Sclater-Booth (President of the Local Government Board) and John Lambert.[78] The deputation included several high-profile manufacturers – Edmund Muspratt, Alfred Allhusen, Hugh Lee Pattinson, Charles Wigg, Alexander Chance and Henry Wade Deacon, to press their case for more caution with extending legislation while accepting many of the recommendations from the Royal Commission. For Sclater-Booth and Lambert the meeting was an opportunity to assess the strength of support on particular issues before any revised legislation was put in hand.

On 7 April 1879 a Bill was introduced into Parliament containing all the main recommendations in the Royal Commission's report and included a greater role for local authorities in their application of the Public Health Act of 1875 in dealing with nuisances. Unfortunately the Bill did not proceed through its second reading because of threatened opposition, and a further attempt the following year was not successful because a general election was called that resulted in a new President of the Local Government Board, J.C. Dodson (who later became Lord Monk-Bretton). Lambert seemed to develop a less cooperative relationship with Dodson and passed the drafting of the new legislation to Angus Smith.[79] This gave Angus Smith the opportunity to take an approach that, while embracing the main recommendation of the Royal Commission and key elements of the last two abortive bills, could omit some parts that irritated the manufacturers, such as the proposed role of local authorities in taking legal action on nuisances. Angus Smith and his inspectors were very clear that the revised legislation should focus on regulating the acid gases rather than the processes that would alleviate much administration. Angus Smith was very disappointed to find that even with support from campaign groups such as the Association for Prevention of Noxious Vapours (Manchester and East Cheshire) and the National Association for the Promotion of Social Science, the clause was missing from the legislation when it was submitted to Parliament on 7 February 1881. As the legislation was proceeding through the Commons, Angus Smith was asked to draft the appropriate clause, which he did. It then came as an even greater disappointment to read the final text of the Alkali, etc., Works Regulation Act and find the clause omitted for reasons that remain unclear. Use of 'et cetera' in the title of the Bill was important for the legislation's role in regulating other sectors, and not just the alkali industry. As Roy MacLeod has pointed out, the legislation increased control,

---

[77]   Lambert later acquiesced on the travel grant, although Smith's estimate of £74 was reduced to a grant of £60 by Lambert without explanation. On the salary increase, see letter from Alfred Fletcher to John Lambert dated 20 January 1880. NA (Ref: MH 16/1).

[78]   See report of meeting, *The Times*, 6 January 1879.

[79]   MacLeod, 'Alkali Acts Administration', p. 106.

first, by extending the fixed standards to sulphuric acid and nitric acid; second, by placing more than a dozen new kinds of works under the 'best practicable means' test pending the formulation of fixed standards; and third, by giving the Local Government Board power to expand inspectorial control over cement and salt works by provisional order, as soon as suitable means of regulation could be devised.[80]

Interestingly, the legislation also promoted Angus Smith to Chief Inspector and authorized five more sub-inspectors to assist with the added responsibilities inherent in the terms of the Act. As if to indicate a more permanent place in government responsibilities, salary increases were at last approved for Angus Smith (from £800 to £1000) and for Fletcher, Blatherwick and Todd, the surviving sub-inspectors from the first tranche. Later in 1884 the Provisional Order Clause for cement works and salt works passed without resistance but sadly Angus Smith did not see this success, for he died two days before the Order reached the House of Commons. Within a short period Alfred Fletcher, Angus Smith's loyal colleague in the Inspectorate, was appointed to replace him as Chief Inspector.

By 1884 the legislation and its enforcement through the Alkali Inspectorate had matured to a position where it was difficult to see them not having a permanent place in government. In large measure the working principles were in place and the framework existed for extending regulation to other gases and processes as industry advanced in its ever more sophisticated manner into the twentieth century and beyond. The legislation and the conduct of enforcement by the Alkali Inspectorate had advanced significantly since the tentative steps laid out in the 1863 Act, and Angus Smith had had the vision and determination and had shown outstanding leadership in advancing the regulation of pollution.[81] While air quality was not as good as most wanted, it was certainly a great deal better than it would have been without Robert Angus Smith.

---

[80]   Ibid., p. 107.

[81]   An interesting comparison with the Factory Inspectorate is provided in P.W.J. Bartrip and P.T. Fenn, 'The Evolution of Regulatory Style in the Nineteenth Century British Factory Inspectorate', *Journal of Law and Society*, 10/2 (1883), pp. 201–22.

Chapter 8

Robert Angus Smith's Legacy

The Alkali Inspectorate's Achievements under Robert Angus Smith

In making any assessment of Robert Angus Smith's success as Inspector of the Alkali Inspectorate, it is important to gauge it against the backdrop of the situation that existed in July 1863 when the Royal Assent was given to the Alkali Act. This Act proved to be the first legislation to allow the incursion or interference of government into the workings of industry, and with it a change in government approach from *laissez-faire* to interventionist. The Treasury was resistant to paying realistic salaries for the inspection team; indeed it questioned the number of inspectors needed to provide coverage of the whole country. When Angus Smith was appointed Inspector there was no precedent for the inspection regime's need to enforce the tight regulations that formed part of the legislation. It would be imperative to find a satisfactory working arrangement with manufacturers, who had, during the hearings of the Select Committee on Injury from Noxious Vapours in 1862, frequently expressed opposition to inspectors entering their premises and interfering with the operation of their plant. When he took up the post of Inspector Angus Smith must have felt a great weight of responsibility, given the very challenging and daunting situation lying ahead.

Whatever the detailed shortcomings in the legislative framework and its enforcement, the Alkali Inspectorate was deemed successful from quite early in its existence. As the various chapters of this book have shown, as the Alkali Inspectorate demonstrated the competence and ability to fulfil all the responsibilities placed upon it, its remit was enhanced to widen and tighten the regulation of chemical processes beyond just works operating the Leblanc process. A major part of this success is due to the diligence, commitment and leadership of Robert Angus Smith.

At the start Angus Smith was able to appoint a team of sub-inspectors with the knowledge and experience to ensure that inspections were carried out in the thoroughly scientific manner drawn up by him. But also, and perhaps more important for the future success of the Alkali Inspectorate, they developed the good working relationship with manufacturers and thereby became peripatetic consultants, a role that contributed so much to the development of chemical plant and to the reduction in pollution caused by noxious gases from chemical processes. Alfred Fletcher was a close associate of Angus Smith's and contributed many innovative ideas to the work of the Alkali Inspectorate, whether the development of the automated self-acting aspirator or the proposal to apportion financial damages to the area affected by a particular pollution escape. It was no surprise that Fletcher was appointed Chief Inspector following Angus Smith's death in service in 1884.

What is perhaps more surprising is the lack of any official recognition given to Angus Smith following his death compared with John Simon's resignation from the Medical Department of the Local Government Board, as Roy MacLeod has pointed out:

> His [Smith's] scientific work won its own reward: the Establishment, however, hardly noticed his passing. No remark occurs in the Inspectorate files, no mention in the official reports. It is ironic that his good work did not merit the notice that Simon enjoyed, when the latter resigned from the Medical Department eight years before; Smith's passive nature aroused nothing of the scandal that attended his contemporary's tumultuous career.[1]

It is interesting to record that the most vocal acclamation of Angus Smith's success came from the manufacturers (with the noted exception of Hussey Vivian, the Swansea copper manufacturer), the very people who had expressed the greatest reservations about the 1863 legislation and the role of the inspectors. Even before Angus Smith's death, manufacturers spoke during government inquiries of the benefits that had accrued from the Alkali Inspectorate and the positive part played by the inspectors in improving the workings of the industry. The manufacturers asked for even more inspectors to make sure that the legislation worked more effectively and across the whole country. But it was also Angus Smith's personal qualities that stood out, as an article in *Chemical News* in 1879 makes clear:

> His thorough knowledge of industrial chemistry, his suggestive and fruitful mind, his tact, courtesy and patience, his clear oversight of the whole field of operations, and of the rival interest to be dealt with, cannot be too highly esteemed … His methods of leading instead of driving the chemical manufacturers, and of substituting wherever possible, remonstrance and advice for summons and fines is perhaps slow, but it is sure.[2]

It is this relationship with manufacturers that has had such an enduring place in the *modus operandi* of the Alkali Inspectorate and probably the reason why its administrative approach and structure continued through until the 1990s. This remains the principal legacy of Robert Angus Smith.

In reviewing the enforcement role of the Alkali Inspectorate, the historian Christine Garwood has asked whether the Inspectorate acted as 'Green Crusaders' or 'Captives of Industry'.[3] As she recognizes, the work of the Inspectorate in its

---

[1]   MacLeod, 'Alkali Acts Administration', pp. 109–10.

[2]   *Chemical News*, 40 (1879), p. 304.

[3]   Christine Garwood, 'Green Crusaders or Captives of Industry? The British Alkali Inspectorate and the Ethics of Environmental Decision Making, 1864–95', *Annals of*

relation to industry, government, landowners, the public and the environment was complex; it was never focused on one area at a time. It is possible to interpret the Inspectorate's work as varying over time to reflect changing circumstances in industry, the firmer enforcement demanded by government, increasing diversification of works under regulation and a greater concern for health towards the end of the nineteenth century.

There were perhaps three phases. The first phase from 1864 to the late 1870s was concerned with getting the Alkali Inspectorate operational and with ensuring that Leblanc process alkali works achieved the minimum regulation level of acid gas to protect the environment while allowing industry to function.[4] This was a period when few of the alkali works (or any chemical works) employed a qualified chemist, which led to the sub-inspectors acting as peripatetic consultants. The second phase during the 1880s saw a wider range of chemical works under regulation but the onus of enforcement was placed on the manufacturers to find the 'best practical means' to keep polluting materials under control; they were employing qualified chemists who could undertake such development work. This proved a very useful approach when the final, agreed method for control of pollutants was still subject to review. The third phase, probably beginning in the 1880s but coming to fruition in the 1890s, was the greater concern with health, both of the workers and of the public.

Robert Angus Smith's Library

Robert Angus Smith built a considerable library of 3,946 books over many years. Rarely do whole libraries of scientists survive intact since books are more often disposed of in an *ad hoc* fashion or are distributed among relatives and friends when the person dies. The collection that became known as the Robert Angus Smith Library reflects his professional work as well as his personal interests. Unlike many scientists who have had institutional affiliations, Angus Smith worked for only a short time at the Royal Manchester Institution after moving to Manchester and thereafter worked as a consulting chemist who needed to rely on his own reference library to support his professional work. Although his laboratory was separate from his home, it is very likely that all his books were brought together on his death to form one collection; this allows an assessment to be made of those books that informed his professional work, together with books that reflected his wider intellectual and cultural interests, which were many.

When Angus Smith died, a group of friends that probably included James Joule and Edward Schunck formed a committee to raise funds to purchase the library so that it was not lost to the city and would act as a permanent memorial of his

---

Science, 61 (2004), pp. 99–117.

[4] This approach was a direct consequence of the 1863 Select Committee on Injury from Noxious Vapours.

service to Manchester. Having purchased the library intact, presumably from the executors of Angus Smith's estate, the committee approached Owens College about acquiring the collection and the College readily accepted. Angus Smith had shared many of the aims of the institution; indeed he had applied for the professorship of chemistry there on two occasions. The committee's only condition stipulated that the books were stamped with a distinguishing mark as forming part of the Robert Angus Smith Library, and in due course a bookplate: 'Manchester Owens Library/Presented by/The Angus Smith Memorial Committee/1884', was placed in each book. During the formal presentation at Owens College on 15 January 1885, Alfred Neild accepted the gift on behalf of the Governors, and the Principal, Dr Greenwood, spoke of the importance of the collection:

> while the college library had received by donation no fewer than 15,000 volumes, this was, with the exception of the library bequeathed by the late Bishop Lee, the largest single collection presented to it, as it was certainly one of the most valuable.[5]

Although the College had intended to include the Library in its planned printed catalogue, this was never completed, although some individual parts were completed and printed. Thereafter the books lost their unifying origination and were scattered across the classification system. The Library now forms part of the University of Manchester Library following several organizational changes: Owens College became an institutional part of Victoria University (1880), Victoria University of Manchester (1903) and the University of Manchester (2004) after amalgamation with the University of Manchester Institute of Science and Technology (UMIST). Some books from the Robert Angus Smith Library have passed into the Special Collections at the John Rylands Library, Deansgate, which is part of the University of Manchester Library.

Unfortunately, it is very difficult to locate individual books from the Robert Angus Smith Library within the current university library.[6] The Library's original entry in the Owens College accession register dated 5 June 1885 records the addition of 3,946 volumes but does not list them individually. However, there was a separate summary of the books under broad headings:

---

[5]   *The Manchester Guardian*, 16 January 1885.

[6]   Those sections of the catalogue that were completed can be used to identify books from the Robert Angus Smith Library because the entries carry the original accession number assigned to the collection as a whole (35210) in the notes field. However, the current university online catalogue does not allow searches on the notes field, making it necessary to trawl through the card index catalogue that pre-dates the online catalogue.

Atlases 3	Natural philosophy 129
Fine arts 7	Chemistry 358
Theology 150	Metallurgy 105
Philology (general) 133	Natural history 39
Philology (classical) 213	Medicine 97
Mental and moral sciences 95	Poetry 125
Political sciences 26	Novels 99
Sanitary 210	Drama 15
History (general) 244	Miscellaneous 136
History (Scottish) 333	Parliamentary papers 22
History (Irish) 60	Alkali papers 29
History (Celtic) 84	Bibliographical 5
Occult science 89	Italian Statistical Society 83
Mixed science 122	Guidebooks 47

The surviving printed catalogues of the Owens College Library reveal interesting details within their subject area. Besides the expected dictionaries and grammar books in English, German and French, Angus Smith also had them in Spanish, Italian, Dutch and Scandinavian languages (including Icelandic). These no doubt helped with his wide reading for professional purposes, but were available as an aid when either reading novels in several of these languages or for his researches in Celtic history with their links to myths, ballads and songs in related cultures.

It is difficult to summarize adequately and concisely such an extensive library, but I have tried to identify groups of books that might have informed Angus Smith's work in its different forms, writing articles, giving talks, preparing submissions to parliamentary inquiries or associated with his work as Chief Inspector of the Alkali Inspectorate. An example is provided by the topic of air and ventilation, which was important for Angus Smith in relation to his work for the Royal Commission on Mines (1864) and for his work on domestic air quality in which ventilation played such an important part in creating a healthy living environment.[7] His collection of books on air and ventilation included:

Bernan, Walter, *On the history and art of warming and ventilating rooms and buildings*, 1845
Hartley, Walter N., *Air and its relation to life*, 1875
Leeds, Lewis W., *A treatise on ventilation*, 1876
Moore, Ralph, *Ventilation of mines*, 1859
Morin, Arthur J., *Études sur la ventilation*, 1863
Péclet, Eugène, *Traité elementaire de physique*,1847
Perkins, E.E., *A Practical treatise on gas and ventilation*, 1856

---

[7] Ventilation formed a key element of several of his articles for the MLPS, including 'On Some Physiological Effects of Carbonic Acid and Ventilation', *Memoirs of the MLPS*, 3 (3rd series) (1868), pp. 91–107.

Reid, David B., *Illustrations of the theory and practice of ventilation*, 1844
Tyndall, John, *Essays on the floating-matter of the air*, 1881

It is interesting to note that Angus Smith was always keen to have access to books with the latest ideas and would readily seek out foreign sources. The missing periodicals would have allowed an even closer scrutiny of the development of his ideas.

Another important subject area for Angus Smith was broadly meteorology and water; this is reflected in another group of his books that included:

Buchan, Alexander, *Handy book of meteorology*, 1868
Dalton, John, *Meteorological observations and essays*, 1834
Daubeny, Charles, *On climate*, 1863
Drew, John, *Practical meteorology*, 1860
Forster, Thomas, *Researches about atmospheric phaenomena*, 1813
Harrison, James B., *Some observations on the contamination of water by the poison of lead*, 1852
Lowe, Edward J., *Natural phenomena and chronology of the seasons*, 1870
Rowell, George, *An essay on the cause of rain and its allied phenomena*, 1859
Saussure, Horace de, *Essais sur l'hygrométrie*, 1783

As expected, Angus Smith's library contained a rich and diverse collection of books on different aspects of chemistry – theoretical, practical, analytical and as applied to different arts and industries – by well-known authors from Britain (including associates in Manchester), continental Europe or the United States – including books likely to have influenced Angus Smith during his formative years. These included:

Bergman, Torbern, *Chemical Essays*, 1791
Gmelin, Johann, *Geschichte der Chemie* 1797
Lavoisier, Antoine, *Traité élémentaire de chimie*, 1789
Roscoe, Henry, and Carl Schorlemmer, *Chemistry, Volume 1,2,3*, 1877–79
Schorlemmer, Carl, *Rise and development of organic chemistry*, 1879
Stahl, Georg, *Philosophical Principles of Universal Chemistry*, 1730
Thomson, Thomas, *System of Chemistry (Volume 1, 2, 3, 4)*, 1802

Books relating to industrial or applied chemistry were important for Angus Smith in his work for the Alkali Inspectorate and in his consultancy work, and included:

Agricola, Georg, *De re metallica*, 1559
Anderson, Thomas, *Elements of Agricultural Chemistry*, 1860
Bessemer, Henry, *On a New System of Manufacturing Sugar from the Cane*, 1852
Crookes, William, *A practical handbook on dyeing and calico printing*, 1874

Harrison, James B., *Some remarks on the contamination of water by the poison of lead; and its effects on the human body,* 1852

Lunge, Georg, *The Manufacture of Sulphuric Acid and Alkali, Volume 1,2,3,* 1880

Odling, William, *Lectures on animal chemistry,* 1866

Pasteur, Louis, *Études sur la bière,* 1876

Prideaux, Thomas S., *On the economy of fuel,* 1853

Rennie, Robert, *Essays on the natural history and origin of peat moss,* 1807

Smith, Henry A., *The chemistry of sulphuric acid manufacture,* 1873

Chemical dictionaries were well represented, including those by Watts and Nicholson. Books on chemical analysis were important to Angus Smith in his professional work and included:

Bunsen, Robert, *Tafeln zur berechnung gasometrischer analysen,* 1857

Jarman, George, *Qualitative analysis,* 1870

Liebig, Justus von, *Handbook of organic analysis,* 1853

Slater, John, W., *Handbook of chemical analysis for practical men,* 1861

Wanklyn, James A., *Milk analysis,* 1874

Wöhler, Friedrich, *Handbook of inorganic analysis,* 1859

Books with intriguing titles that reflect the increasing pervasiveness of chemistry included:

Benays, Albert J., *Household chemistry,* 1852

Griffiths, Thomas, *Chemistry of the four seasons,* 1846

Piesse, Septimus, *The art of perfumery,* 1855

Youmans, Edward, L., *A handbook of household science,* 1860

Late in his life Angus Smith expressed the view that a prominent scientific body should take responsibility for the publication of reprints of important and influential books and papers, but nothing came of this proposal at the time. In his preface to *Bibliotheca Chemica*, the catalogue of James Young's extensive collection of chemical writings, John Ferguson, Professor of Chemistry at the University of Glasgow, suggested that Angus Smith's idea led to the Alembic Club reprints.[8]

The extensive theology section of the library includes books that have been catalogued under the headings – Dogmatic, Pastoral and Controversial Theology, etc. – and reflect the anguish Angus Smith probably felt throughout his life as to whether he should have become a church minister rather than pursuing his interests in chemistry. As the selected list below shows, many of the books will have influenced his thinking during the period up to 1840 when he decided to

---

[8]   John Ferguson, *Bibliotheca Chemica: a catalogue of the alchemical, chemical and pharmaceutical books in the collection of the late James Young of Kelly and Durris* (London, 1954).

study in Giessen with Liebig, but even after this period his mind was occupied from time to time by further reflection on this haunting matter.

Beard, J.R., *Historical and artistic illustration of the Trinity*, 1846
Calderwood, H., *The relations for science and religion*, 1846
Clarke, S., *Discourse concerning the Being and attribution of God*, 1823
Crewdson, I., *Beacon to the Society of Friends* (2nd edn), 1835
Leone Abbate, *The Jesuit conspiracy*, 1848
Muggleton, L. and J. Reeves, *A divine looking glass*, 1661
Penn, W. *No cross, no crown*, 1842
Slacxk, C.T., *The vehicle of future knowledge or a compendium system of astrology*, 1794
Swedenborg, E., *A compendium of the theological and spiritual writings*, 1854
Tafel, J.F.L., *Documents concerning the life and Character of E. Swedenborg*, 1841
Taylor, J., *Holy living and dying*, 1850
Train, J., *The Buchanites from first to last*, 1846

As someone who had studied classics from a young age and retained a lifelong interest in the subject, Angus Smith had works by many Roman and Greek authors, including Caesar, Cicero, Euripides, Herodotus, Hippocrates, Lucretius, Ovid, Plato, Sophocles, Tacitus, Thucydides, Virgil and Zenophon.

He had a considerable collection of novels, including works by Daniel Defoe, Benjamin Disraeli, Sir Walter Scott, Tobias George Smollett, Laurence Sterne and Jonathan Swift, but he was a particularly avid collector of poetry; his favourites included Thomas Lovell Beddoes, Lord Byron, Elizabeth Cookson, Ebenezer Elliott (The Corn Law Rhymer), Janet Hamilton, James Hogg, John Hopkins, Henry Wadsworth Longfellow and William Shakespeare.[9]

Unfortunately, among all the information about the library little or no mention is made of the periodicals Angus Smith used. The scientific periodicals were probably stored within his laboratory accommodation where they were close at hand for reference during his scientific investigations and other professional work. Periodicals relating to his other interests were probably stored at his home. However, when Smith died and the Library Committee was appointed, it is quite likely that the periodicals were used to fill any gaps in the holding of Owens College and any remaining ones destroyed. Nevertheless, a list of periodicals would have allowed better insight into how Angus Smith kept up with the latest knowledge and ideas, and the extent to which he was aware of comments and criticisms on and of his work, particularly in his role as Chief Inspector of the Alkali Inspectorate.

---

[9]   See *Class II Languages and Literature: General Teutonic and English Language and Literature*. Owens College Library Catalogues (Manchester, 1899), in the possession of the Archives and Record Centre, The University of Manchester Library.

The Alkali Inspectorate with Alfred Fletcher as Chief Inspector

Following Robert Angus Smith's death in service in May 1884, Alfred Fletcher, as expected, was appointed Chief Inspector. Fletcher was part of the first group of sub-inspectors appointed by Angus Smith in 1864. He had been a loyal and innovative colleague during the initial 20 years when the Alkali Inspectorate was in its formative stage. Fletcher was often the first person to whom Angus Smith referred any outstanding issues or sought advice before reaching a final decision. Fletcher's working background before joining the Alkali Inspectorate provided the necessary experience and insight into the workings of industry that were to form the basis of the peripatetic relationships with manufacturers that proved so crucial in advancing the work of the Inspectorate.

Fletcher was born in 1826 into a Nonconformist family and educated in Berlin before becoming a railway surveyor.[10] However, he soon realized that his real interest was in science and he attended University College London, where he was to receive the Gold Medal in 1851 (the runner-up was Henry Roscoe).[11] Soon afterwards he set up a company to exploit his work on new aniline dyes, but the initiative was stifled by prolonged litigation and he readily accepted Angus Smith's offer of a sub-inspector post within the Alkali Inspectorate in February 1864.[12] As a sub-inspector, Fletcher was given responsibility for the western division, which included Lancashire, Cheshire and Flintshire, and the post was based in Liverpool. The area had a profusion of alkali works that made it one of the key centres alongside Tyneside and Glasgow for the Leblanc process, whose acid gas initiated the accusations of nuisance back in the 1820s. It was Fletcher who had developed the self-acting aspirator for collecting gas samples from flues over several days without the intervention of an inspector; this provided evidence of any accidental gas escapes.

With his appointment as Chief Inspector, Fletcher was to return to his work on the control of black smoke from coal burning, and would resume, perhaps with more force and determination, several of the proposals for more effective regulation that Angus Smith had struggled unsuccessfully to place on the statute book during the last few years of his life as his health failed. These included the important principles of focusing on individual noxious gases rather than the processes in which they originated and the need for manufacturers to adopt the 'best practicable means' as a benchmark for the emission of any noxious gas rather than rely on specific emission limits that quickly became out of date due to advances in process technology.[13]

---

[10] 'Obituary of Alfred Fletcher', *Nature*, 106 (1920), p. 185.

[11] Maisie Fletcher, *The Bright Countenance: A Personal Biography of Walter Morley Fletcher* (London, 1957), p. 18.

[12] Eric Ashby and Mary Anderson, *The Politics of Clean Air* (Oxford, 1981), p. 65.

[13] Ibid., p. 66.

As Ashby and Anderson have covered in detail, almost from his first annual report, Fletcher, with the support of the manufacturers and various campaign groups, sought to persuade his civil service colleagues in the Local Government Board that the principles he wanted enshrined in legislation would bring the regulation of noxious gases up to date and avoid the need for regular amendments to the Acts. The scale of the release of hydrogen sulphide from sulphur waste was becoming a scandal because of the inactivity on the part of the Inspectorate. However, the attention of the President and Secretary of the Board was focused on other matters and they defended their inaction by expressing the view that 1890 was too soon after the 1881 legislation for major amendments.[14] While Angus Smith in his latter years as Chief Inspector would have backed off, Fletcher merely saw this intransigence (perhaps even bloody-mindedness) as deserving of even greater outspokenness at every opportunity until his points were accepted. While the Board's interest in submitting revised legislation ebbed and flowed, by 1891 Fletcher decided to use his annual report for 1890 in another attempt to gain support for the changes that would strengthen the work of the Inspectorate. Finally a revised Bill reached the statute book on 27 June 1892, but unfortunately it was a severe compromise, still addressing processes rather than noxious gases. It was inevitable that further refinement of the legislation would be needed to provide the flexible approach envisaged by Fletcher, who wanted to avoid the regular update in inspector's powers.

Like Angus Smith, Fletcher did not believe in using the legislation as a means to impede the progress of industry and was at pains to avoid using prosecution to enforce the regulations. Unlike Angus Smith, he sought public support for his approach; one way he did this was to engage more prominently in the controversy of black smoke: he was not averse to making reference to this in sections of his annual report as Chief Inspector, even though the problem was still outside the remit of the Alkali Inspectorate. Some progress was made in reducing black smoke in London following the initiatives by Lord Palmerston, but there was a sharp contrast with the excessive smoke found in the industrial towns of the north. Fletcher's annual report of 1886 drew attention to this disparity and the need for more effective burning of coal and the use of smokeless fuels to prevent environmental conditions that were having a worse effect on health than the noxious emissions from chemical processes. His general agitation prompted further interest and action among campaign groups, including the Manchester and Salford Noxious Vapours Abatement Association, which set up an executive committee to oversee a series of tests of appliances to investigate smoke reduction.[15] Putting aside a possible conflict of interest, Fletcher accepted the position of Chairman. The committee was strong, not just with do-gooders but more importantly with some knowledgeable engineers who understood the scientific as well as the technical issues. Although trials were carried out over the period 1891–96, little in the way

---

14   Ibid., pp. 66–9.
15   Ibid., p. 73.

of conclusive evidence was obtained that allowed any initiatives to move forward, even though Fletcher had taken some personal action and installed a very efficient heating system in his own house that reduced running costs considerably.[16]

Fletcher retired as Chief Inspector in May 1895; over the 11 years in the post he had subjected his masters within the Local Government Board to a constant barrage of demands to improve the workings of the regulatory process. He certainly brought greater energy and determination to bear and sought the support of the manufacturers as well as the public. His vision and efforts kept many of the ideas outlined initially by Angus Smith at the forefront of the pollution control debate.

Evolution of the Alkali Inspectorate to the 1990s

It was left to Fletcher's successor as Chief Inspector, R.F. Carpenter, working in 1906 with the Board's President, John Burns, to bring forward legislation that incorporated a clause on the 'best practicable means' to control a list of regulated noxious gases rather than depend on a list of processes; this avoided the need for regular updates of the regulations as chemical processes proliferated.[17] With the inclusion of other key amendments as well, the 1906 Act at last contained those elements of regulation sought over a long period of time by Angus Smith and Alfred Fletcher, and was to provide the regulatory framework for the Inspectorate's work to at least 1975, if not beyond to the 1990s.

By 1956, the Alkali Inspectorate was responsible for 1,794 processes operating in 921 works in England and Wales, and for 116 processes in 82 works in Scotland. As if to demonstrate the continuing need to review chemical processes and update regulation, in 1956 W.A. Damon (Chief Inspector from 1929 to 1955) raised concerns over a number of industries and processes, including sulphuric acid processes, viscose processes, cement manufacture and petroleum refining, while reviews were taking place on electricity-generating stations, coal carbonization works, ironworks, steelworks, ceramic works and fluorine emissions.[18] Although new processes or gases might raise concerns as the chemical industry proliferated, the regulatory framework was now in place whereby the industry could speedily be brought under the regulation of the Alkali Inspectorate. This was due, in the main, to the pioneering and diligent work of Robert Angus Smith and then his successor Alfred Fletcher as chief inspectors.

The Standing Royal Commission on Environmental Pollution was set up in 1970 and in 1974 the Commission was asked to review the administrative structure regarding air pollution control in view of the Alkali Act (1863) and the Clean Air Act (1956), and any potential conflicts that might arise.[19] Its report of January 1976

---

16 Ibid., p. 74.
17 Ibid., pp. 79–81.
18 Damon, 'The Alkali Act', pp. 568–73.
19 Ashby and Anderson, *Clean Air*, p. 128.

recommended that work associated with the two elements of legislation should be combined under the responsibility of one body, and in addition recommended that the new body should be known as Her Majesty's Pollution Inspectorate (HMPI). The HMPI would apply its considerable knowledge and experience to resolving major pollution problems, while advising other bodies such as local authorities and water authorities.[20]

During the Inspectorate's 150-year existence, responsibility for it had passed between many different government departments and agencies, and its name has changed on more than one occasion: Board of Trade, 1863—72; Local Government Board, Alkali Inspectorate, 1873–1918; Ministry of Health, Alkali Inspectorate, 1919—51; the Chief Inspector's independence disappeared when the Inspectorate was transferred to the Health and Safety Executive in 1975; it was known as the Industrial Air Pollution Inspectorate from 1983 to 1987 and became Her Majesty's Inspectorate of Pollution (HMIP) when it was transferred back to the Department of the Environment in 1987. Since 1 April 1996 HMIP has been part of the Environment Agency.

Alleviation of 'Black Smoke' and Smog

As we saw earlier, it was a pall of black smoke hanging over Manchester that first alerted Angus Smith in 1844 to the effect of this pollution on the well-being of the town and its people. During his evidence to the Royal Commission on Noxious Vapours on 3 August 1876, Angus Smith, when asked about the likely amount of coal being burnt in the country for all purposes, replied 'about 110 million tons'; this figure would rise very steadily over the years while coal remained the principal source of energy. Black smoke had been a source of pollution almost from the earliest days of its introduction in the twelfth century. Attempts to reduce the level of smoke by more efficient burning, using better stoves and hearths, were largely unsuccessful. While during the nineteenth century local authorities had powers to enforce smoke reduction, little use was made of them; many of the elected representatives had vested interests in the factories and workshops producing much of the smoke. As we have seen, even Angus Smith and Alfred Fletcher (even more prominently) in their role as Chief Inspector were unable to harness sufficient interest and concern about black smoke to cause any major change either in the legislative framework or in the practices of industry and in domestic settings.[21]

When heavy clouds of black smoke are accompanied by dense fog it is called smog. The term is usually applied in cases where fog is very dense and it is barely possible to see beyond arm's reach. The fog descends to low levels under the

---

20   Ibid., p. 129.

21   Carlos Flick, 'The Movement for Smoke Abatement in 19th-Century Britain', *Technology and Culture*, 21/1 (1980), pp. 29–50.

influence of certain meteorological conditions so that the cloud remains almost stationary over several days, causing damage to public health. Smogs in Britain usually contain black smoke, vehicle exhaust fumes and sulphur dioxide (or sulphurous acid). The presence of tarry materials (as in the 1952 episode) gives smog its characteristic yellow tinge; then the smog is known as a 'pea-souper'.[22]

Major incidents of smog were reported in London during 1873, 1880 and 1905, though the impact on public health is not well recorded.[23] With the frequency of such episodes, the government began to accept their occurrence much as they accepted flooding, as natural events that had to be endured because it was powerless to act. While London County Council took some initiatives to reduce black smoke in the early years of the twentieth century, these had little impact on the overall level of smoke across the city; similar inaction in the northern industrial towns resulted in similar major smog incidents and damage to health.

A further major smog episode occurred in London in December 1952. While the government barely acknowledged it, newspaper reports recorded transport delays and cancellations; sporting events cancelled or abandoned as players 'disappear into the gloom'; a performance of *La Traviata* at Sadler's Wells Theatre abandoned after the first act because of the audience's inability to see the performers; and the masking of criminal activities under cover of the smog. Air quality monitoring equipment failed to activate due to the level of pollution and size of particulate matter in the smog (about 3000 $\mu g/m^3$, compared to about $30\mu g/m^3$ or less today).[24]

Even though an initial estimate of mortalities was put at 4,000, the government showed no interest in taking any action and certainly not in new legislation. Several government ministers came under intense questioning in Parliament but still they barely acknowledged that there was a problem, let alone the need to take action. By May 1953, and under intense pressure, the government at last agreed to set up a committee of inquiry, but it was only just before the committee first met in July 1953 that Sir Hugh Beaver was announced as the chairman. The committee included several specialists in pollution control who were aware of a similar

---

[22]   It has been suggested that the yellow tinge is due to the smoke particles absorbing the blue wavelength of the sunlight above the smog. For technical details of air pollution, see R.S. Scorer, 'Technical Aspects of Air Pollution', in Allan D. McKnight, Pauline K. Marstrand and T. Craig Sinclair (eds), *Environmental Pollution Control: Technical, Economic and Legal Aspects* (London, 1974), pp. 43–62.

[23]   Mark Z. Jacobson, *Atmospheric Pollution: History, Science and Regulation* (Cambridge, 2002), pp. 85–6; Peter Brimblecombe, 'Air Pollution and Health History', in Stephen, T. Holgate, Jonathan M. Samet, Hillel S. Koren and Robert L. Maynard (eds), *Air Pollution and Health* (London, 1999), pp. 11–16. Peter Brimblecombe has included some statistics for 1873 and 1880 in his book, *The Big Smoke* (London, 1987), p. 124.

[24]   Devra L. Davis, Michelle L. Bell and Tony Fletcher, 'A Look Back to the London Smog of 1952 and the Half Century Since', *Environmental Health Perspectives*, 110/12 (December 2002), p. 734.

incident in the industrial town of Donora, Pennsylvania in 1948. The full report of the committee was published in November 1954 to positive reviews; even the government gave an approval in principle that would inevitably allow some room and time for re-evaluation and perhaps even for the report to gather dust on a shelf. But as Ashby and Anderson have contended, it was the tone of the report and its recommendations that would not allow the government to avoid taking action:

> We wish to state our emphatic belief that air pollution on the scale with which we are familiar in this country today is a social and economic evil which should no longer be tolerated, and that it needs to be combated with the same conviction and energy as were applied one hundred years ago in securing pure water. We are convinced that, given the will, it can be prevented. To do this will require a national effort and will entail costs and sacrifices.[25]

Another key event took place while the Beaver Committee was sitting: the City of London initiated parliamentary legislation in March 1954 to create smokeless zones. Besides bringing the issues into parliamentary discussion again, it warned the government (and especially its key ministers) that legislative inaction over pollution from black smoke would not be tolerated. It is interesting to speculate whether even at this stage the government would have avoided legislation, but in November 1954, with the Beaver Committee report imminent, the government's hand was forced when the Tory MP for Kidderminster, Sir Gerald Nabarro, won the vote to present a private member's bill to Parliament. Sir Gerald was a flamboyant character known for pursuing rather eccentric policies. Nevertheless, when his bill was debated in Parliament it received such wide support among members, although some felt it should go further, that the government could see that it would be voted through. The government at this point admitted defeat and agreed to bring its own bill to the House if the private bill was withdrawn; this Nabarro promptly did, sensing victory for common sense and the wish of Parliament.

The government bill was given its first reading on 26 July, some five months after withdrawal of Nabarro's bill, and from the outset was a watered-down version taking account of the interest of parties with vested interests such as industry, local authorities and the civil service. While many felt that the powers of the Act should rest with the Alkali Inspectorate, in the end they came under the control of the local authorities, given their ability to create smoke-free zones. Industrial interests were keen to reduce as far as possible any unreasonable demands and increased costs. Nevertheless, the final Clean Air Bill received the royal assent on 5 July 1956. This proved to be one of the most distinguished episodes in parliamentary history, when the weight of support from ordinary MPs forced an obdurate government into legislative action.[26]

---

[25]   Ashby and Anderson, *Clean Air*, pp. 106–7.
[26]   Ibid., pp. 111–13.

There has been a debate over the years about the part played by the Clean Air Act in reducing the occurrences of smog and thick fogs. The number of such episodes was dramatically reduced over quite a short period but other factors contributed to this outcome, such as the switch from coal to electricity and gas for heating, the adoption of central heating in homes, and the movement of people from the city and town centres to the suburbs. These factors added to the effectiveness of the Clean Air Act, and, had they not taken place, there is every likelihood that the Act would have brought about major change, albeit more slowly. Fogs today in Britain have become meteorological phenomena without the added menace of heavy black smoke and a far cry from the experience of Robert Angus Smith walking into Manchester on a December afternoon in 1844.

Much more recently researchers have reviewed the statistical information associated with the 1952 episode in attempting to re-evaluate the number of deaths. Initial government figures gave the mortalities due to the smog episode at about 4,000, but later it was noted that a high level of mortalities continued for the months through to the end of March. A government report in 1954 attributed these later mortalities to an influenza epidemic that was concurrent with the smog. The recent re-evaluation reviewed the scale of influenza epidemic that would probably cause 8,000 deaths and concluded 'that such an epidemic would have to be about three times larger than the most severe epidemics recorded from 1949 to 1968'.[27] One study concludes with some justification that 'if the excess deaths in the months after the 1952 London smog are related to air pollution, the mortality count would be approximately 12,000 rather than the 3,000 to 4,000 generally reported for the episode'.[28]

It is interesting to note that smogs associated with Los Angeles have a different etiology; they are caused by the motor car's inefficient burning of petrochemical products, resulting in an aerosol composed of polymerized oxidation products of unsaturated hydrocarbons, part of a photochemical cycle involving the oxides of nitrogen.[29] A further contributory factor is temperature inversion caused by subsistence of descending air that warms when it is compressed over cool, moist air. Los Angeles started its pollution controls in 1945. Arie Jan Haagen-Smit, a Dutch-born scientist working at Caltech, whose specialist interest had been the micro-chemistry of natural products (pineapples in fact), started applying these micro-chemical techniques to the study of smogs. He soon found that the inefficient working of the motor car engine was responsible for the harmful aerosols. Having

---

[27]   Michelle Bell, Devra L. Davis and Tony Fletcher, 'A Retrospective Assessment of Mortality from the London Smog Episode of 1952: The Role of Influenza and Pollution', *Environmental Health Perspectives*, 112/1 (January 2004), p. 8.

[28]   Ibid.

[29]   The elucidation of the mechanisms involved in the formation of these aerosols is due to the work of Arie Jan Haagen-Smit (1900–1977), who lobbied for the establishment of California's Air Resources Board, becoming its Chairman in 1968. See James Bonner, *Arie Jan Haagen-Smit (1900–1977): A Biographical Memoir* (Washington, 1989).

become Chairman of California's Air Resources Board in 1968 he took on the US automobile industry, and in 1969 even had the sale of Volkswagen cars banned until the firm was prepared to comply with the exhaust emission levels determined by the Board.

The removal of smog episodes in Britain should not be allowed to hide the serious harm from air pollution at the current time. An air quality report by the House of Commons Environmental Audit Committee in 2010 drew attention to the government's failure to meet many of the European targets for safe air pollution with stark statistics:

> The report highlighted that life-expectancy was reduced on average by 7–8 months because of poor air quality, while in the worst affected areas that could have been as high as 9 years. Research suggested that between 30,000 and 50,000 people a year were dying prematurely because of it. Air pollution was also causing significant damage to ecosystems.[30]

The main concern is limits for nitrogen dioxide (NO_2 and particulates. Aphekom, a European research project funded by the European Commission, concluded:

> Those living near main roads in cities could account for some 15–30 per cent of all new cases of asthma in children and chronic obstructive pulmonary disease and coronary heart disease in adults 65 years of age and older.[31]

Other research has not shown a link between external air pollution with asthma. In his book *Allergy: The History of a Modern Malady*, Mark Jackson has pointed out that

> comparative studies charting the prevalence of asthma in Germany following the opening and removal of the Berlin Wall in 1989 and subsequent political reunification in 1990 showed paradoxically that allergies were less common in heavily polluted East German cities than in their West German counterparts. Likewise, several other cities or regions with high levels of atmospheric pollution, such as Athens in Greece and parts of China, generally experienced low levels of asthma. By contrast, some countries with relatively clean air, such as Scotland and New Zealand, demonstrated high rates of allergic diseases during the second half of the twentieth century.[32]

While public suspicions concentrate on outdoor air pollution, more recent research on allergies has begun to focus on the influence of the indoor environment, where a number of pollutants are present such as dust, organic solvents, cleaning agents,

---

[30]   *Environmental Audit Committee, Fifth Report of Session 2009-10, Air Quality*, HC 229.

[31]   Ibid., p. 7.

[32]   Mark Jackson, *Allegy: The History of a Modern Malady* (London, 2006), p. 164.

furnishing materials and finishes. Conclusive evidence is still awaited and even though air quality has improved substantially since 1952, there is no room for complacency.

The Continuing Challenge of 'Acid Rain'

When Robert Angus Smith first investigated acid rain he focused on coal with its sulphur content as the principal cause. However, coal is not the only culprit. There are other natural sources as well as human activities that need to be taken into account. The principal natural source is volcanoes, during which high concentrations of sulphur dioxide are produced, resulting in low acidity (pH values of about 2). Ever since Angus Smith used the term 'acid rain' for the first time in 1859, human activities have been the principal cause for the increasing occurrence of acid rain, in terms of both the quantities released and distribution that crosses national boundaries. To combat the effects of acid rain national governments must not only take action within their own jurisdictions but also enter into transnational and even intercontinental agreements, since acid rain recognizes no national boundaries.

Since 1859 there has been a dramatic, ongoing increase in the quantities of coal consumed, often coal with a high content of sulphur, as industrial expansion has continued on an ever larger and global scale. But other fossil fuels have also emerged, such as oil used for power generation and fuels for motor cars. Attention has focused increasingly on efficient burning of fuels in the internal combustion engine to reduce the nitrogen oxides as well as sulphur dioxide in the emissions. Besides contributing to acid rain, the internal combustion engine can contribute to photochemical smogs in specific meteorological conditions. Development and wider use of electric vehicles will help to ameliorate the damage from these sources. From time to time, as research and closer monitoring have continued, other sources of sulphur dioxide have emerged; these include ocean traffic around the world as ships tend to rely on cheaper fuel oil, which has a very high sulphur content (up to 5 per cent).[33] Also, extensive algae masses have been shown to release the chemical dimethyl sulphide, which is oxidized in the air to form sulphur dioxide; recent research has shown that 'in summer 25 per cent of the acid rain falling over Europe now comes from huge algae masses in the North Sea'.[34]

What is the effect of acid rain? Acid rain has been shown to affect large swathes of forests, freshwater lakes (and their fish stocks) and soils. As Angus Smith noted, forests and woodlands can be dramatically affected by acid rain, with foliage stripped. The soil can be changed with the action of acid rain whereby microbes are unable to function and some metal ions are leached out, affecting the nutrients upon which the trees rely. Many freshwater lakes have suffered the

---

[33]   John Ashton and Ron Laura, *The Perils of Progress* (London, 1998), p. 8.
[34]   Ibid., p. 35.

effects of acid rain and attempts have been made to mitigate this by applying lime. This has proved very harmful to the ecology of the lakes and the animal life they support.[35] This situation has prompted further research to find a solution to the problem. Rivers have also been affected, resulting in damage to the spawning of fish where lower pH values (and therefore greater acidity) reduce the biodiversity and the ability of some species of fish to thrive or to even hatch.[36] Overall, the simple fact remains that every effort must be made to reduce the levels of acid rain at source rather than tinkering with the effects at the margins.

The production and distribution of acid gases have become global phenomena closely associated with industrial development and expansion, especially in recent times with the emerging economies spread throughout the world, particularly in Asia. Some of the worst examples of acid rain occurred in the 1970s in Soviet bloc countries where inefficient industrial plant relied on poor-quality coal containing a high percentage of sulphur as energy source. Little investment was made to improve the burning of this coal and little attempt made to search out new sources of energy. To disperse the smoke and acid gases, the approach in the UK and in most places worldwide has been to use very tall smoke stacks that protected the areas in the immediate vicinity of the factory plant but caused the emissions to be dispersed over much larger areas, cutting across national boundaries. Much of the acid rain generated in the former Soviet bloc countries ravaged extensive forestry areas in Scandinavia to the north. The emissions from electricity-generating plants in the UK were also found to contribute to the damage inflicted on these forests.

North America has not been immune to acid gases, with both Canada and the United States experiencing the impact of acid rain on freshwater and terrestrial ecosystems, buildings and health. In 1980 the US Congress approved the Acid Deposition Act, which established the National Acidic Precipitation Assessment Program (NAPAP) and started an 18-year research and assessment programme to understand how best to approach the challenges of acid rain. Although hampered by internecine disputes, Congress in 1989 passed amendments to the Clean Air Act. One of these amendments set control limits on sulphur dioxide and nitrogen oxides, with notably a reduction of 10 million tons in releases of sulphur dioxide. Increasingly, acid rain from one state affected other states. Accordingly, in 2005 the Environmental Protection Agency issued the Clean Air Interstate Rule (CAIR) with targets set across the eastern states and the District of Columbia to reduce 2003 levels of sulphur dioxide by over 70 per cent and nitrogen oxides by over 60 per cent. As the EPA has reported:

> In New England, between 1990 and 2000, we have seen a 25% decrease in NO_x emissions from all sources (from approximately 897,000 tons to 668,000 tons).

---

[35]   S. Woodin and U. Skiba, 'Liming fails the acid test', *New Scientist*, 10 March 1990, pp. 30–34.

[36]   United States Environmental Protection Agency website (www.epa.gov/acidrain), and see 'Effects of Acid Rain – Surface Waters and Aquatic Animals'.

Between 2000 and 2006, NO_x emissions from Acid Rain by power plants in New England have further decreased by more than 31,000 tons. During that same period, SO_2 emissions from those power plants have decreased by 54% (from approximately 211,000 tons to 96, 500 tons).[37]

In Europe, governmental action was taken earlier when in 1979 the Geneva Convention on Long-range Transboundary Air Pollution was established by the United Nations Economic Commission for Europe. This was a belated response to the finding of researchers of a causal link between sulphur dioxide emissions in continental Europe and acidification of Scandinavian lakes in the 1960s.

> [This] resulted in the signature of the Convention on Long-range Transboundary Air Pollution by 34 Governments and the European Community (EC). The Convention was the first international legally binding instrument to deal with problems of air pollution on a broad regional basis. Besides laying down the general principles of international cooperation for air pollution abatement, the Convention sets up an institutional framework bringing together research and policy.[38]

The Convention covers the entire UNECE region (with special focus on Eastern Europe, the Caucasus and Central Asia and South-East Europe) and shares its knowledge and information with other regions of the world. It will be interesting to see how the global emerging economies such a China, India and Brazil heed the experiences of countries that have already suffered the impact of acid rain.

What would Robert Angus Smith make of the current situation if he were living today? He would fully appreciate the importance of these issues and how the problem has grown on an even wider scale. But he would be very surprised, and probably shocked and dismayed, that more has not been done to reduce the occurrence of acid rain and its damaging impact. He would certainly be a leading figure in the fight to get the origins and impact of acid rain more widely understood, but more importantly, he would ensure that action was taken at the international level in order to reduce the damage to the natural environment as well as the dangers to human health.

---

[37]   United States Environmental Protection Agency website (www.epa.gov/acidrain) and see 'History of Acid Rain'.

[38]   United Nations Economic Commission for Europe website (www.unec.org/enu/lrtap/) and see 'Convention'.

Chapter 9
Epilogue

Robert Angus Smith's name is not readily recognized today even by scientists, for it is not associated with a new theory or new chemical reaction as a contribution to chemical understanding. However, he is one of the first scientists who should be recognized for the contribution he made in applying scientific understanding and knowledge to the field of sanitary science, a field to emerge during his lifetime.

By studying the many factors that influence air and water quality, Angus Smith ensured that such ideas not only contributed to the continuing debate within scientific circles but also retained a high profile among those influencing and formulating government policies. It was inevitable during the formative period of a relatively new field to find that experimental investigations and results were often interpreted as little more than isolated and fragmented science-based information without a clear thesis or hypothesis binding them together and thereby adding to our corpus of knowledge. As Edward Schunck recorded of Angus Smith's work:

> At the time when Dr. Smith started his researches, sanitary science could not be said to exist, unless a mere collection of unconnected facts can be dignified with the name of science. Since that time much more system has been introduced into the subject, and a great portion of the merit of having developed the purely scientific side of it is due to Dr. Smith.[1]

This work took place during a period when there was urgent need for better scientific understanding to bring about a change in the living conditions of a large proportion of society.

Angus Smith never sought the limelight or any honour for the work he did. He was always generous with his time and the results of his investigations, readily collaborating with others and sharing results. This attitude is shown in his work for the Manchester Literary and Philosophical Society and the Manchester and Salford Sanitary Association. But he also showed this quality in his numerous contributions to government inquiries and in the relationship he sought with manufacturers through his work with the Alkali Inspectorate.

It is perhaps as the first Inspector (and later Chief Inspector) of the Alkali Inspectorate that Robert Angus Smith's name will be best remembered. Having found a blank slate for the workings of the Inspectorate and the inspection procedures, he put in place working principles such as the use of chemical analysis techniques and adoption of the 'best practical means' approach that transformed the enforcement's effectiveness, to a level that manufacturers were gradually

---

[1]   Schunck, 'Angus Smith', p. 94.

convinced of the merits of legislation that many had initially opposed. Such principles provided the underpinning to the workings of the Alkali Inspectorate and its successor bodies until the 1990s.

In reviewing his lifetime's work in sanitary science it is possible to recognize that Angus Smith may never have completely lost his aspiration to become a minister of the Church. His spiritual outlook was probably never far away in his work as a public servant and many might accuse him of proselytizing a spiritual outlook for society:

> To keep the air in our towns fresh and wholesome, to restore the water of our streams to its pristine clearness, this is the kind of work to which [Angus] Smith dedicated his life, and at which he laboured to the very last.[2]

In promoting wider benefits for society, Angus Smith shared the aspiration and determination of his long-time mentor, Justus von Liebig, but it was the commitment to these ideals that many others, as civil scientists, were to take up following Angus Smith's death and that remain such an important and potent part of Angus Smith's legacy. The 150th anniversary of the establishment of the Alkali Inspectorate is on 1 January 2014; that the Inspectorate is still functioning, albeit under a different name, is a further testament to the important part Robert Angus Smith played in laying the foundation for regulating air pollution in Britain.

---

2   T.E. Thorpe, 'Robert Angus Smith', *Nature,* 30 (1884), p. 104.

Robert Angus Smith Bibliography

Manuscripts

Balliol College, Oxford: copy of letter from Florence Nightingale to Angus Smith (1865).

British Library, London: correspondence with Florence Nightingale (1866) (Add. 45799).

Cheshire and Chester Archives and Local Studies Centre, Chester: LNWR Crewe, Register of Water Analysis (1853–83).

Florence Nightingale Museum, London: correspondence with Florence Nightingale (1854–66).

National Archives, London: correspondence relating to Alkali Administration with Board of Trade, Local Government Board and Treasury (1863–84).

Royal Society, London: correspondence and documents, including Certificate of Election (1857–82).

University College London: correspondence with Edwin Chadwick (1845–84) in Special Collections Library.

Books and Reports

Air and Rain: The Beginnings of a Chemical Climatology (London, 1872).

Chemical and Physical Researches of Thomas Graham (Edinburgh, 1976).

Descriptive List of Antiquaries near Loch Etive (Glasgow, 1870).

Disinfectants and Disinfection (Edinburgh, 1869).

Ellida Vatn and Kjalarnes in Iceland (Edinburgh, 1874).

Intermediate Report of the Chief Inspector, 1863 and 1874, of his proceedings since the passage of the latter Act, P.P. 1876 (165), xvi.

The Life and Works of Thomas Graham (Glasgow, 1884).

Loch Etive and the Sons of Ulsnach (London, 1879).

Loch Etive and the Sons of Ulsnach. With illustrations by Miss J. Knox Smith (London, 1885).

Memoir of Dr. Dalton, and History of Atomic Theory (Manchester, 1856).

'On a Method of Estimating Carbonic Acid in the Air', *Report of the British Association for the Advancement of Science Meeting 1865 (Sect.)* (London, 1866), pp. 35–7.

'On Sulphuric Acid in the Air and Water of Towns', *Report of the British Association for the Advancement of Science Meeting 1851, (Pt. 2)*, (London, 1852), p. 52.

'On the Air and Water of Towns', *Report of the British Association for the Advancement of Science Meeting 1848* (London, 1849), pp. 16–31.

'Report on the Air of Mines', in *Report of Royal Commission of Mines*, P.P. 1864 (3389), XXIV.

'Report to the Cattle Plague Commission', 25 April 1866, in *Third Report of the Commissioners appointed to enquire into the origin and nature, etc., of the cattle plague, and an appendix*, P.P. 1866 (3656), pp. 155–86.

To Iceland in a Yacht (Edinburgh, 1873).

Visit to St. Kilda in the 'The Nyanza' (Glasgow, 1879).

With Frederick Crace-Calvert

Report: Messrs. Spence and Dixon's Alum Works. Observations etc. (Manchester, 1855).

With E. Schunck and H.E. Roscoe

'On the Recent Progress and Present Condition of Manufacturing Chemistry in the S. Lancashire District', *Report of the British Association for the Advancement of Science Meeting 1861* (London, 1862), pp. 108–28.

Journal Articles[1]

Abrasion of Iron Rails on Railways, *Proceedings of the MLPS*, 5 (1866), p. 23.

Absorption of Gases by Charcoal, *Chemical News*, 7 (1863), pp. 242–3.

Absorption of Gases by Charcoal, Part II. On a New Series of Equivalents or Molecules, *Proceedings of the Royal Society*, 28 (1879), pp. 322–4.

The Air of Houses and Workshops, *Transactions of the National Association for the Promotion of Social Science, Sheffield Meeting 1865–1866* (London: John W. Parker, 1867), pp. 419–27.

Ancient Maps of Africa, *Proceedings of the MLPS*, 3 (1864), pp. 229–30.

Archaeology of the Voice: Scottish Lowlands, *Proceedings of the Society of Antiquaries of Scotland* XVI (1881–82) pp. 451–57.

Armstrong's Gun, *Proceedings of the MLPS*, 1 (1860), p. 108.

Arsenic in Coal Pyrites and in the Atmosphere, *Proceedings of the MLPS*, 2 (1862), p. 5.

Cattle Plague and Use of Carbolic Acid, *Proceedings of the MLPS*, 5 (1866), pp. 118–19.

---

[1]   The *Proceedings of the MLPS* report meetings of the Society and discussions at meetings, while the *Memoirs of the MLPS* contain the full papers given at the Society's meetings.

A Centenary of Science in Manchester, For the 100th Year of the Literary and Philosophical Society of Manchester, *Memoirs of the MLPS*, 9 (3rd series) (1883), pp. 1–487.

Chemical Climatology, *Journal of the Scottish Meteorological Society*, 3 (1873), pp. 2–11.

Composition of the Atmosphere, *Proceedings of the MLPS*, 4 (1864), pp. 30–32.

Crystallised Amalgam of Sodium, *Proceedings of the MLPS*, 3 (1864), pp. 106–7.

Dancer's Aspirator, *Proceedings of the MLPS*, 4 (1865), pp. 26–7.

Decomposition of Phosphate of Lime, *Proceedings of the MLPS*, 3 (1864), p. 107.

Derivation and Composition of Rosolic Acid, *Proceedings of the MLPS*, 1 (1860), pp. 13–14.

Description of a Meteorite which Fell at Allport in Derbyshire, *Memoirs of the MLPS*, 9 (2nd series) (1851), pp. 146–8.

Descriptive List of Antiquities near Loch Etive, *Proceedings of the Society of Antiquaries of Scotland* IX (1870–72) pp. 396–418.

Descriptive List of Antiquities near Loch Etive, Argyllshire, Consisting of Vitrified Forts, Cairns, Crannogs, etc.: With Some Remarks on the Growth of Peat, *Proceedings of the Society of Antiquaries of Scotland* IX (1870–72) pp. 81–106.

Descriptive List of Antiquities near Loch Etive. Part III Continued, *Proceedings of the Society of Antiquaries of Scotland* X (1872–74) pp. 70–90.

Descriptive List of Antiquities near Loch Etive. Part IV, *Proceedings of the the Society of Antiquaries of Scotland* XI (1874–76) pp. 298–305.

Descriptive List of Antiquities near Loch Etive. No: 5. Plan of Dun-Mac Uisneachan, *Proceedings of the the Society of Antiquaries of Scotland* XII (1876–78) pp. 13–19.

Detection of Fire-damp, *Chemical News*, 39 (1879), pp. 851–9.

The Distribution of Ammonia, *Memoirs of the MLPS*, 6 (3rd series) (1879), pp. 267–78.

The Distribution of Ammonia, *Proceedings of the MLPS*, 17 (1878), pp. 188–93.

Effects of Dews and Fogs in Producing Epidemics, *Proceedings of the MLPS*, 5 (1866), pp. 22–3.

Electrical Experiments in the Thames, *Proceedings of the MLPS*, 1 (1860), p. 66.

The *Eucalyptus* Near Rome, *Proceedings of the MLPS*, 15 (1876), pp. 150–61.

The *Eucalyptus* Near Rome, *Memoirs of the MLPS*, 6 (3rd series) (1879), pp. 25–36.

An Examination into the Products of the Putrefaction of Blood, *Proceedings of the MLPS*, 2 (1862), pp. 127–9 and 241–4.

An Examination into the Products of the Putrefaction of Blood, *Memoirs of the MLPS*, 2 (3rd series) (1865), pp. 47–63.

The Examination of Air, *Proceedings of the Royal Society*, 26 (1878), pp. 512–17.

Examples of Relative and Absolute Law, *Proceedings of the MLPS*, 2 (1862), pp. 147–9.

Lead in Manchester Water, *Proceedings of the MLPS*, 2 (1862), pp. 30–32.

Limits of Chemical and Mechanical Action, *Proceedings of the MLPS*, 2 (1862), pp. 164–5.

Measurement of the Actinism of the Sun's Rays and of Daylight, *Proceedings of the Royal Society*, 30 (1880), pp. 355–9.

Memoir of John Dalton and History of the Atomic Theory up to his Time, *Memoirs of the MLPS*, 13 (2nd series) (1856), pp. 1–298.

Minimetric Method of Analysis, *Proceedings of the MLPS*, 4 (1865), pp. 159–62.

Minimetric Method of Analysis, *Memoirs of the MLPS*, 3 (3rd series) (1868), pp. 187–203.

The Mud of the Clyde, *Proceedings of the Glasgow Philosophical Society*, 12 (1880), pp. 321–8.

Note on the Development of Living Germs in Water, *Proceedings of the MLPS*, 22 (1883), pp. 25–32.

Note on the Word 'Chemia', *Proceedings of the MLPS*, 20 (1881), p. 15.

Notes of Stone Circles in Durries, Kincardineshire and its Neighbourhood, *Proceedings of the the Society of Antiquaries of Scotland* XIV (1879–80) pp. 294–309.

On a Method of Estimating Carbonic Acid in the Air, *Report of the Thirty-Fifth Meeting of the BAAS, Birmingham 1865* (London, 1866), p. 35.

On a Mode of Rendering Substances Incombustible, *Philosophical Magazine*, 24 (1849), pp. 116–19.

On a Peculiar Fog Seen in Iceland and on Vesicular Vapour, *Memoirs of the MLPS*, 5 (3rd series) (1876), pp. 150–64.

On a Remarkable Fog in Iceland, *Proceedings of the MLPS*, 12 (1873), p. 11.

On a Vitrified Mass of Stone from the Fort of Glen Nevis, *Proceedings of the MLPS*, 22 (1883), pp. 1–3.

On Air from off the Atlantic and from Some London Law Courts, *Proceedings of the MLPS*, 5 (1865), pp.115–16.

On Air from the Mid-Atlantic, and from Some London Law Courts, *Memoirs of the MLPS*, 3 (3rd series) (1868), pp. 181–6.

On an Apparatus for Collecting the Gases from Water and Other Liquids, and its Application in General Chemical Analysis, *Memoirs of the MLPS*, 12 (2nd series) (1855), pp. 271–5.

On Causes Preventing Smoothness of Ground, *Memoirs of the MLPS*, 5 (3rd series) (1876), pp. 223–7.

On Exhibitions on the Continent, *Proceedings of the MLPS*, 7 (1868), pp. 28–32.

On Organic Matter in the Air, *Proceedings of the MLPS*, 9 (1870), pp. 65–76.

On Putrefaction of Blood, *Memoirs of the MLPS*, 2 (3nd series) (1865), pp. 47–62.

On Sewage and Sewage Rivers, *Memoirs of the MLPS*, 12 (2nd series) (1855), pp. 155–76.

On Some Physiological Effects of Carbonic Acid and Ventilation, *Proceedings of the MLPS*, 4 (1865), pp. 79–83.

On Some Physiological Effects of Carbonic Acid and Ventilation, *Memoirs of the MLPS*, 3 (3rd series) (1868), pp. 91–107.

On Some Ruins at Ellida Vatn and Kjalarnes in Iceland, *Proceedings of the the Society of Antiquaries of Scotland* X (1872–74) pp. 151–177.

On Sulphuric Acid in the Air and Water of Towns, *Report of the Twenty-First Meeting of the BAAS, Ipswich 1851* (London, 1851), p. 52.

On the Absorption of Gases by Charcoal, *Proceedings of the MLPS*, 6 (1867), pp. 195–6.

On the Absorption of Gases by Charcoal, *Report of the Thirty-Eighth Meeting of the BAAS, Norwich 1868* (London, 1869), pp. 44–5.

On the Action of Town Atmospheres on Building Stone, *Proceedings of the MLPS*, 12 (1873), pp. 19–20.

On the Air of Towns, *Journal of the Chemical Society*, 9 (1859), pp. 196–235.

On the Air and Rain of Manchester, *Memoirs of the MLPS*, 10 (2nd series) (1852), pp. 207–18.

On an Apparatus for Collecting the Gases from Water and Other Liquids, and its Application in General Chemical Analysis, *Memoirs of the MLPS*, 12 (2nd series) (1855), pp. 271–6.

On the Composition and Derivation of Rosolic Acid, *Memoirs of the MLPS*, 15 (2nd series) (1860), pp. 1–7.

On the Composition of Atmospheric Air and Rain-water, *Journal of the Chemical Society*, 10 (1872), pp. 33–4.

On the Composition of the Atmosphere, *Proceedings of the MLPS*, 4 (1865), pp. 30–32.

On the Composition of the Atmosphere, *Memoirs of the MLPS*, 3 (3rd series) (1868), pp. 1–55.

On the Development of Living Germs in Water, *Proceedings of the MLPS*, 22 (1883), pp. 25–32.

On the Estimation of the Organic Matter of the Air, *Proceedings of the Royal Institution*, 3 (1858–62), pp. 89–94.

On the Examination of Air, *Proceedings of the Glasgow Philosophical Society*, 7 (1871), pp. 326–30.

On the Examination of Water for Organic Matter, *Proceedings of the MLPS*, 7 (1868), pp. 72 and 78–9.

On the Examination of Water for Organic Matter, *Memoirs of the MLPS*, 4 (3rd series) (1871), pp. 37–88.

On the History of the Word 'Chemistry' or 'Chemia', *Proceedings of the MLPS*, 19 (1880), pp. 141–7.

On the Meteorological Instruments Invented by Dr. Joule, *Proceedings of the MLPS*, 4 (1865), pp. 132–3.

On the Production and Prevention of Malaria, *Memoirs of the MLPS*, 1 (3rd series) (1862), pp. 222–33.

On the 'Spark Tube', or Inflammable Gas Indicator, *Transactions of the Manchester Geological Society*, 15 (1880), pp. 384–6.

On the Water of the River and Firth of Clyde, *Proceedings of the Glasgow Philosophical Society*, 6 (1868), pp. 154–63.

On Vitrified Forts, *Proceedings of the MLPS*, 13 (1874), pp. 19–20.

On Water from Peat and Soil, *Memoirs of the MLPS*, 8 (2nd series) (1848), pp. 377–80.

Optical Phenomena, *Proceedings of the MLPS*, 3 (1864), pp. 39–40.

Production and Prevention of Malaria, *Proceedings of the MLPS*, 2 (1862), pp. 44–7.

Products of the Putrefaction of Blood, *Proceedings of the MLPS*, 2 (1862), pp. 127–9 and 241–4.

Relation of Work and Workers, *Proceedings of the MLPS*, 2 (1862), pp. 47–9.

Report to the Local Government Board as to Treatment of Sewage, *Chemical News*, 41 (1880), pp. 50–52.

Science in Our Courts of Law, *Journal of the Society of Arts*, 7 (18 Nov. 1859–16 Nov. 1860), pp. 136–7.

A Search for Solid Bodies in the Atmosphere, *Proceedings of the MLPS*, 7 (1868), pp. 154–7.

A Search for Solid Bodies in the Atmosphere, *Memoirs of the MLPS*, 4 (3rd series) (1871), pp. 266–70.

Sewage, *Proceedings of the MLPS*, 1 (1860), p. 79.

The Smoke Question, *Chemical News*, 14 (1866), pp. 182–5.

Some Remarks on the Air and Water of Towns, *Philosophical Magazine*, 30 (1847), pp. 478–82.

Some Ancient and Modern Ideas of Sanitary Economy, *Memoirs of the MLPS*, 11 (2nd series) (1854), pp. 39–90.

A Study of Peat, Part 1, *Proceedings of the MLPS*, 14 (1875), pp. 117–22.

A Study of Peat, Part 1, *Memoirs of the MLPS*, 5 (3rd series) (1876), pp. 281–345.

Sur la composition de l'air des mines du Cornouailles, *Cuyper, Revue Univ.*, 20 (1866), pp. 79–89.

Sur la Condensation des Vapeurs Acides Dans les Fabriques de Soude, *Moniteur Scientifique*, 21 (1879), pp. 851–59.

Study of Peat, Pt. 1, *Memoirs of the MLPS*, 5 (3rd series) (1876), pp. 281–345.

Ventilation and the Reason for It, *Popular Science Monthly*, 1 (1872), pp. 356–62.

Who are the Celts, *Proceedings of the Society of Antiquaries of Scotland* XVII (1882–83) pp. 385–99.

The Word 'Chemia' or Chemistry, Memoirs *of the MLPS*, 7 (3rd series) (1882), pp. 101–25.

General Bibliography

Parliamentary Debates, Papers and Statutes

Alkali Act 1863, P.P. 1863 (135).
Alkali Act (1863) Amendment Act, P.P. 1874 (99).
Alkali Act (1863) Perpetuation Act 1868, P.P. 1867-68 (153)
Eleventh Report of the Inspector under the Alkali Act, P.P. 1876 (C.1339).
Environmental Audit Committee, Fifth Report of Session 2009–10, Air Quality, HC 229.
First Annual Report of the Registrar-General, Appendix (P), P.P. 1839 (187).
First Report of the Inspector appointed under the Alkali Act, P.P. 1865 (3460).
Fourth Report of the Inspector appointed under the Alkali Act, P.P. 1867–68 (3988).
Health of Towns Commission, Reports of Commissioners, Appendix, Part II, Report on the Sanitary Condition of the Large Towns of Lancashire, P.P. 1845 (602) (601),
Ninth Report of the Inspector appointed under the Alkali Act, P.P. 1874 (C.815).
Public Health Act 1872, P.P. 1872 (261).
Report of Royal Commission of Mines, P.P. 1864 (3389), XXIV.
Report of the Select Committee on Injury from Noxious Vapours, P.P. 1862 (486), xiv.
Royal Commission on Noxious Vapours. I. Report, P.P. 1878 (XLIV).
Royal Commission on Noxious Vapours. II. Minutes of Evidence, P.P. 1878 (XLIV).
Second Report of the Inspector appointed under the Alkali Act, P.P. 1866 (3701).
Third Report of the Commissioners appointed to enquire into the origin and nature, etc., of the cattle plague, and an appendix, P.P. 1866 (3656).
Third Report of the Inspector appointed under the Alkali Act, P.P. 1867 (3792).
Twenty-First Annual Report of the Alkali Inspector, P.P. 1885 (C. 4461).

Newspapers and Magazines

Liverpool Journal
Liverpool Mercury
The Daily Post
The Guardian
Manchester Guardian
The Manchester Guardian
The Mirror
Punch
The Times

Manuscript Sources

Wellcome Library, London: Florence Nightingale correspondence, Ref: Ms. 8999/23.
Liverpool Record Office, Liverpool: Derby Collection, correspondence with Florence Nightingale.

Archive Sources

Catalyst Science Discovery Centre, Widnes: transcript of Diary of Peter Spence.
Cheshire and Chester Archives and Local Studies Centre, Chester: United Alkali Co. Ltd., Alkali Manufacturers Association 1876–1881.
Greater Manchester County Record Office, Manchester: Records of the Manchester and Salford Sanitary Association.
Lancashire Record Office, Preston: Rowson and Cross Papers.
Liverpool Record Office, Liverpool: Muspratt Papers.
National Archives, London: correspondence relating to Alkali Administration with Board of Trade, Local Government Board and Treasury (1863 to 1884).
Tyne and Wear Archives, Newcastle upon Tyne: records of Newcastle Town Council.
Worcestershire County Record Office, Worcester: Records of the Stoke Prior works of the British Alkali Company.

Books, Theses and Journal Articles

A Full Report of the Trial of the Important Indictment Preferred by the Corporation of Liverpool Against James Muspratt, Esq., Manufacturer of Alkali at the Liverpool Spring Assizes 1838 before Sir John Taylor Coleridge, Knight, and Special Jury, for a Nuisance Alleged to Proceed from his Chemical Works in Vauxhall Road, Liverpool (Liverpool: D. Marples, 1838). [Copy in Special Collections, Harold Cohen Library, University of Liverpool.]

Alford, W.A.L., and J.W. Parkes, 'Sir James Murray: a pioneer in the making of superphosphate', Chemistry and Industry, 1953, p. 852.

Allen, J. Fenwick, Some Founders of the Chemical Industry (London: Sherratt and Hughes, 1906).

André, George, Spon's Encyclopaedia of the Industrial Arts, Manufactures and Commercial Products (London: E. and F. Spon, 1879).

Anon, 'Air Pollution by Chemical Works', Quarterly Journal of Science, 7 (1870), pp. 330–41.

Anon, 'The British Association for the Advancement of Science', Chemical News (11 October 1862), pp. 189–90.

Anon, 'Dr. Angus Smith, FRS', The Biograph and Review, 5 and 6 (1879–81), pp. 142–52.

Anon, 'Edward Schunck', Journal of the Society of Chemical Industry, 50 (1931), p. 65.

Anon, 'Obituary: Robert Angus Smith, LL.D, F.R.S.', Journal of the Society of Chemical Industry, 3 (1884), pp. 316–17.

Anon, 'Obituary: Robert Angus Smith', American Journal of Science, 128 (1884), pp. 79–80.

Anon, 'Obituary: Robert Angus Smith', Proceedings of the Society of Antiquaries of Scotland XIX (1884–85) p. 6.

Anon, 'Report of the Committee on Scientific Evidence in Courts of Law', Report of the Nottingham Meeting of the British Association for the Advancement of Science in 1866 (London, 1867), pp. 456–7.

Anon, 'Robert Angus Smith', The British Medical Journal, 1 (1884), p. 976.

Anon, 'Robert Angus Smith', Journal of Chemical Society, 47 (1885), pp. 335–7.

Anon, 'Scientific Assessors in Courts of Justice', Nature, 38 (1888), pp. 289–91.

Anon, Testimonials in favour of Robt. Angus Smith, Ph.D., F.R.S., F.C.S., candidate for the Chair of Chemistry in the University of Aberdeen (London, 1862).

Anon, 'The Whole Duty of a Chemist', Nature, 33 (1885), pp. 73–7.

Ashby, Eric, and Mary Anderson, The Politics of Clean Air (Oxford: Clarendon Press, 1981).

Ashenburg, Katherine, The Dirt on Clean: An Unsanitized History (New York: Farrar, Straus and Giroux, 2007).

Ashton, John and Ron Laura, The Perils of Progress (London: Zen Books Ltd, 1998).

Ashton, Thomas S., Economic and Social Investigations in Manchester 1833–1933 (Brighton: Harvester Press, 1977).

Axon, W.E.A., Cobden as a Citizen (Manchester: Fisher Unwin, 1907).

Baker, T., 'On the Plan Suggested by the Government Commissioners for Disposing of the Metropolitan Sewage', Journal of Society of Arts, 6 (1858), pp. 416–23.

Barker, T.C., 'Lancashire Coal, Cheshire Salt and the Rise of Liverpool', Transactions of the Historic Society of Lancashire and Cheshire (1951), pp. 83–101.

Bartrip, P.W.J., 'British Government Inspection, 1832–1875: Some Observations', The Historical Journal, 25 (1982), pp. 605–26.

Bartrip, P.W.J., and P.T. Fenn, 'The Evolution of Regulatory Style in the Nineteenth Century British Factory Inspectorate', Journal of Law and Society, 10/2 (1883), pp. 201–22.

Bebbington, David, 'Gospel and Culture in Victorian Nonconformity', in Jane Shaw and Alan Kreider (eds), Culture and the Nonconformist Tradition (Cardiff: University of Wales Press, 1999).

Beck, Ann, 'Some Aspects of the History of Anti-Pollution Legislation in England, 1819–1954', Journal of the History of Medicine, 14 (1959), pp. 475–89.

Bell, Michelle, Devra L. Davis and Tony Fletcher, 'A Retrospective Assessment of Mortality from the London Smog Episode of 1952: The Role of Influenza and Pollution', Environmental Health Perspectives, 112/1(January 2004), pp. 6–8.

Bonner, James, Arie Jan Haagen-Smit (1900–1977): A Biographical Memoir (Washington: National Academy of Sciences, 1989).

Brimblecombe, Peter, The Big Smoke (London: Methuen, 1987).

Brimblecombe, Peter, 'Air Pollution and Health History' in Stephen, T. Holgate, Jonathan M. Samet, Hillel S. Koren and Robert L. Maynard (eds), Air Pollution and Health (London: Academic, 1999).

Broad, D.W., Centennial History of the Liverpool Section Society of Chemical Industry 1881–1981 (London: Society of Chemical Industry, 1981).

Brock, W.H. (ed.), Justus von Liebig und August Wilhelm Hofmann in ihren Briefen (Weinheim: Vertag Chemie, 1984).

Brock, W.H., Justus von Liebig, The Chemical Gatekeeper (Cambridge: Cambridge University Press, 1997).

Brüggemeier, Franz-Josef, Das unendliche Meer der Lüfte: Luftverschmutzung, Industrialisierung und Risikodebatten im 19. Jahrhundert (Essen: Klartext, 1996).

Campbell, Alex, The Chemical Industry (London: Longman, 1971).

Cannadine, David, 'Engineering History, or the History of Engineering? Re-Writing the Technological Past', Transactions of the Newcomen Society, 74 (2004), pp. 163–80.

Chance, Alexander M., 'The Recovery of Sulphur from Alkali Waste by Means of Lime Kiln Gases', Journal of the Society of Chemical Industry, 50 (1931), pp. 151–61.

Clegg, Samuel, A Practical Treatise on Manufacture and Distribution of Coal Gas (London: John Weale, 1841).

Clow, Archibald and Nan L. Clow, The Chemical Revolution (London: Batchworth Press, 1952).

Crace-Calvert, Frederick, 'Manufacture and Properties of Carbolic Acid', The Lancet, (14 December 1867), pp. 733–4.

Crookes, William, 'The Evidence of Experts', Chemical News, V (5 April 1862), p. 281.

Crookes, William, 'Sir William Crookes on Psychic Research', Smithsonian Report, 1898, pp. 185–205. Includes extracts from Crookes's address to the BAAS Meeting in Bristol in 1898 and his address as President to the Society for Psychical Research in 1897.

Damon, W.A., 'The Alkali Act and the Work of the Alkali Inspectors', Royal Society of Health Journal (London), 76/9 (1956), pp. 566–75.

Davis, Devra L., Michelle L. Bell and Tony Fletcher, 'A Look Back to the London Smog of 1952 and the Half Century Since', Environmental Health Perspectives, 110/12 (December 2002), p. 734.

Dickens, Charles, Hard Times (London: Wordsworth, 1995).

Dingle, A.E., '"The Monster Nuisance of All": Landowners, Alkali Manufacturers, and Air Pollution, 1828–64', Economic History Review, 35 (1982), pp. 529–47.

Donnelly, J.F., 'Representations of Applied Science: Academies and Chemical Industry in Late Nineteenth-Century England', Social Studies in Science, 16 (1986), pp. 195–234.

Donnelly, J.F., 'Consultants, Managers, Testing Slaves: Changing Roles for Chemists in the British Alkali Industry, 1850–1920', Technology and Culture, 35 (1994), pp. 100–128.

Douglas, Mary, Purity and Danger: an Analysis of Concepts of Pollution and Taboo (London: Routledge and Kegan Paul, 1966).

Evelyn, John, Fumifugium or The Inconvenience of the Air and Smoke or London Dissipated (London, 1661).

Eyler, John M., 'The Conversion of Angus Smith: The Changing Role of Chemistry and Biology in Sanitary Science, 1850–1880', Bulletin of the History of Medicine, 54 (1980), pp. 216–34.

Ferguson, John, Bibliotheca Chemica: A Catalogue of the Alchemical, Chemical and Pharmaceutical Books in the Collection of the Late James Young of Kelly and Durris (London: Derek Verschoyle Academic and Bibliographical Publications, 1954).

Fletcher, A.E. 'Modern Legislation in Restraint of Emission of Noxious Gases from Manufacturing Operations', Journal of the Society of Chemical Industry, 11 (1892), pp. 120–24.

Fletcher, Maisie, The Bright Countenance, A Personal Biography of Walter Morley Fletcher (London: Hodder and Stoughton, 1957).

Flick, Carlos, 'The Movement for Smoke Abatement in 19th-Century Britain', Technology and Culture, 21/1 (1980), pp. 29–50.

Fournier-d'Albe, E.E., Life of Sir William Crookes (London: T. Fisher Unwin, 1923).

Fressoz, Jean-Baptiste and Thomas Le Roux, 'Protecting the factories and commodifying the environment: the great transformation of French pollution regulation, 1700–1840', in Geneviève Massard-Guilbaud and Stephen Mosley (eds), Common Ground. Integrating the Social and Environmental in History (Newcastle: Cambridge Scholars, 2011), pp. 340–66.

Fullmer, June Z., 'Technology, Chemistry, and the Law in the Early 19th-Century England', Technology and Culture, 21/1 (1980), pp. 1–28.

Garfield, Simon, The Last Journey of William Huskisson: The Day the Railway Came of Age (London: Faber, 2002).

Gibson, Alan, 'Robert Angus Smith and Sanitary Science', MSc Diss., University of Manchester Institute of Science and Technology, 1972.

Gibson, A., and W.V. Farrar, 'Robert Angus Smith, F.R.S., and "Sanitary Science"', Notes and Records of the Royal Society of London, 74 (1974), pp. 241–62.

Gibson, Jack, Pollockshaws: A Brief History (Pollockshaws Heritage Group, 2006).

Gilbert, Pamela K., Mapping the Victorian Social Body (Albany, NY: State University of New York Press, 2004).

Golan, Tal, Laws of Men and Laws of Nature (Cambridge, MA: Harvard University Press, 2004).

Gorham, Eville, 'Robert Angus Smith, F.R.S., and "Chemical Climatology"', Notes and Records of the Royal Society of London, 36/2 (1982), pp. 267–72.

Griffiths, Ralph A., Singleton Abbey and the Vivians of Swansea (Llandysul: Gomer, 1988).

Haber, L.F., The Chemical Industry during the Nineteenth Century (Oxford: Clarendon Press, 1958).

Hamlin, Christopher, 'Scientific Method and Expert Witnessing: Victorian Perspectives on a Modern Problem', Social Studies of Science, 16 (1986), pp. 485–513.

Hamlin, Christopher, Public Health and Social Justice in the Age of Chadwick (Cambridge: Cambridge University Press, 1998).

Hardie, D.W.F., and J. Davidson Pratt, A History of the Modern British Chemical Industry (Oxford: Pergamon Press, 1966).

Henry, William, 'Experiments on the Quantity of Gases absorbed by Water, at different Temperatures, and under different Pressures', Philosophical Transactions, 93 (1803), pp. 29–42.

Hope, E.W., in collaboration with W. Hanna and C.O. Stallybrass, Industrial Hygiene and Medicine (London: Ballière, Tindall and Cox, 1923).

Hudson, John, Chemistry and the British Railway Industry 1830-1923, PhD diss., Open University, 2005.

Hunt, Tristram, Building Jerusalem (London: Weidenfeld and Nicolson, 2004).

Hylton, Stuart, A History of Manchester (Chichester: Phillimore, 2003).

Jackson, Mark, Allergy: The History of a Modern Malady (London: Reaktion Books, 2006).

Jacobson, Mark Z., Atmospheric Pollution: History, Science and Regulation (Cambridge: Cambridge University Press, 2002).

Kargon, Robert H., Science in Victorian Manchester (Manchester: Manchester University Press, 1977).

Kingzett, C.T., The History, Products and Processes of the Alkali Trade (London: Longmans Green and Co., 1877).

Kidd, Alan, Manchester: A History (Lancaster: Carnegie Publishing, 2006).

Laurie, P., 'The Chemical Trades', in Thomas Oliver (ed.), Dangerous Trades (London: John Murray, 1902), pp. 268–98.

Liebig, Justus, 'On Poisons, Contagions, and Miasmas', Report of the Tenth Meeting of the British Association for the Advancement of Science in Glasgow August 1840 (London, 1841), pp. 72–3.

Liebig, Justus, 'On the Azotised Nutritive Principle of Plants', Annals of Electricity, 10 (1843), pp. 483–4.

Liebig, Justus von, An Address to the Agriculturalists of Great Britain (Liverpool, 1845).

Liebig, Justus von, Familiar Letters on Chemistry, 3rd edn (London: Taylor, Walton and Maberly, 1851).

Luckin, Bill, Pollution and Control (Bristol: Hilger, 1986), pp. 164–7.

Lunge, G., A Theoretical and Practical Treatise on the Manufacture of Sulphuric Acid and Alkali, vol. 1 (London: John Van Voort, 1879).

MacLeod, Roy M., 'The Alkali Acts Administration, 1863–84: The Emergence of the Civic Scientist', Victorian Studies, 9 (1965–66), pp. 85–112.

Malthus, Thomas, An Essay on the Principle of Population (London, 1798).

McDonald, Lynn (ed.), Florence Nightingale on Society, Philosophy, Science, Education and Literature (Waterloo, Ontario: Wilfrid Laurier University Press, 2003).

McDonagh, Oliver, 'The Nineteenth Century Revolution in Government: A Reappraisal', The Historical Journal, 1 (1958), pp. 52–67.

McDougall, Ellen, The McDougall Brothers and Sisters (London, 1923).

Meadows, Jack, The Victorian Scientist (London: British Library, 2004).

Medhurst, Richard G., Kathleen M. Goldney and Mary R. Barrington (eds), Crookes and the Spirit World (London: Souvenir, 1972).

Meetham, A.R., Atmospheric Pollution (Oxford: Pergamon Press, 1956).

Morrell, Jack, 'Thomas Thomson: Professor of Chemistry and University Reformer', British Journal for the History of Science, 4 (1969), pp. 245–65.

Morrell, Jack, 'The Chemist Breeders: The Research Schools of Liebig and Thomas Thomson', Ambix, 19 (1972), pp. 1–46.

Morrell, Jack and Arnold Thackray, Gentlemen of Science: The Early Years of the British Association for the Advancement of Science (Oxford: Clarendon Press, 1981).

Mosley, Stephen, The Chimney of the World: A History of Smoke Pollution in Victorian and Edwardian Manchester (Cambridge: White Horse, 2001).

Muspratt, J. Sheridan, Chemistry, Theoretical, Practical and Analytical as Applied and related to Arts and Manufactures (2 vols), (Glasgow: Mackenzie, 1860).

Newell, Edmund, 'Atmospheric Pollution and the British Copper Industry, 1690–1920', Technology and Culture, 38 (1997), pp. 655–89.

Newell, Edmund, '"Copperopolis": The Rise and Fall of the Copper Industry in the Swansea District, 1826–1821', Business History, 32/3 (1990), pp. 75–97.

Noakes, Richard, '"Cranks and Visionaries": Science, Spiritualism and Transgression', PhD diss., University of Cambridge, 1998.

Odling, William, 'Science in Courts of Law', Journal of the Society of Arts, 8 (1860), pp. 167–8.

Oesper, Ralph E., 'Justus von Liebig – Student and Teacher', Journal of Chemical Education, 4/12 (1927), pp. 1461–7.

Oesper, Ralph E., 'Nicholas Leblanc (1742–1806)', Journal of Chemical Education, 19 (1942), pp. 567–72.

Oesper, Ralph E., 'Nicholas Leblanc (1742–1806)', Journal of Chemical Education, 20 (1943), pp. 17–19.

Observations in refutation of certain charges preferred against the Alum Works of Messrs. Spence & Dixon ... With an examination of Dr. Smith and Profr. Calvert's Report on the same (and copy of that Report) (Manchester, 1855).

Pettenkofer, A., 'Volumetric Estimation of Atmospheric Carbonic Acid', Quarterly Journal of the Chemical Society, 10 (1858), pp. 292–7.

Quinn, Kevin, Boulby Alum: The Works Diary of George Dodds 1772–1788. Research Report Number 9 (Cleveland Industrial Archaeology Society), 2011.

Rawson, R.W., 'On the Sulphur Trade of Sicily and the Commercial Relations with that Country and Great Britain,' Journal of the Statistical Society, 2 (1839), p. 449.

Reed, Peter, 'Acid Towers and Weldon Stills in Leblanc Widnes', Journal of the North Western Society for Industrial Archaeology and History, 2 (1977), pp. 3–8.

Reed, Peter, 'Where Even the Birds Cough: The First British Cases of Large-scale Atmospheric Pollution by the Chemical Industry on Merseyside and Clydeside in the Early 19th Century', in M. Fetizon and W.J. Thomas (eds), The Role of Oxygen in Improving Chemical Processes (6th BOC Priestley Conference), (Cambridge: Royal Society of Chemistry, 1993), pp. 115–22.

Reed, Peter, 'Robert Angus Smith and the Alkali Inspectorate', in E. Homburg, A.S. Travis and H.G. Schröter (eds), The Chemical Industry in Europe, 1850–1914: Industrial Growth, Pollution, and Professionalization (Dordrecht: Kluwer Academic Publishers, 1998), pp. 149–63.

Reed, Peter 'Entry for Robert Angus Smith', Biographical Dictionary of 19th Century British Scientists (Bristol: Thoemmes, 2004), pp. 1843–6.

Reed, Peter, 'Acid Towers and the Control of Chemical Pollution 1823–1876', Transactions of the Newcomen Society, 78 (2008), pp. 99–126.

Reed, Peter, 'The Alkali Inspectorate 1874–1906: Pressure for Wider and Tighter Pollution Regulation', Ambix, 59 (2012), pp. 131–51.

Reid, Thomas Wemyss, Memoirs and Correspondence of Lyon Playfair, First Lord Playfair of St. Andrews (London: Cassell and Co., 1899).

'Report of the Committee on Scientific Evidence in Courts of Law', in Report of the Nottingham Meeting of the British Association for the Advancement of Science in 1866 (London, 1867).

Richardson, T., and H. Watt, Chemical Technology, vol. 1 (part 3) (London: Baillière, 1863).

Roderick, Gordon W., and Michael D. Stephens, 'Profits and Pollution: Some Problems facing the Chemical Industry in Liverpool in the Nineteenth Century. The Corporation of Liverpool versus James Muspratt, Alkali Manufacturer, 1838', Industrial Archaeology, 11/2 (1974), pp. 35–45.

Russell, Colin A., and John Hudson, Early Railway Chemistry and its Legacy (Cambridge: Royal Society of Chemistry, 2012).

Russell, Colin A., with Noel G. Coley and Gerrylynn K. Roberts, Chemists by Profession (Milton Keynes: Open University Press, 1977).

Schabas, Margaret, 'Review of "The Rise of Statistical Thinking 1820–1900" (Princeton, 1986)', Victorian Studies, 31 (1987), p. 123.

Schunck, Edward, 'Memoir of Robert Angus Smith', Memoirs of the MLPS, 10 (3rd series) (1887), pp. 90–102.

Scorer, R.S., 'Technical Aspects of Air Pollution' in Allan D. McKnight, Pauline K. Marstrand and T. Craig Sinclair (eds), Environmental Pollution Control: Technical, Economic and Legal Aspects (London: Allen and Unwin, 1974), pp. 43–62.

Sherrard, Robert H., 'The White Slaves of England', Pearson's Magazine, 2 (1896), pp. 48–55.

Simon, John, English Sanitary Institutions (London: Cassell and Co., 1890).

Singer, Charles, The Earliest Chemical Industry (London: Folio Society, 1948).

Skempton, W., and J. Brown, 'John and Edward Troughton', Notes and Records of the Royal Society, 27 (1972–73), pp. 233–62.

Smith, William Anderson, Shepherd Smith, The Universalist (London: Sampson Low and Co., 1892).

Smith, Roger and Brian Wynne, 'Introduction', in Roger Smith and Brian Wynne (eds), Expert Evidence: Interpreting Science and the Law (London: Routledge, 1989), pp. 1–22.

Stanley, Michael, The Chemical Work of Thomas Graham, PhD diss. Open University, 1980.

Taylor, F. Sherwood, A History of Industrial Chemistry (London: Heinemann, 1957).

Tennant, E.W.D., 'Early History of the St. Rollox Chemical Works', Chemistry and Industry, 66 (1947), pp. 667–73.

Thorpe, T.E., 'Robert Angus Smith', Nature, 30 (1884), pp. 104–5.

Thorsheim, Peter J., Inventing Air Pollution: The Social Construction of Smoke in Britain, 1880–1920, PhD diss., University of Wisconsin-Madison, 2000.

Tocqueville, Alexis de, Journeys to England and Ireland, trans. G. Lawrence and K.P. Mayer, (London: Faber and Faber, 1958).

Vallée,Gérard (ed.), Florence Nightingale on Health in India (Waterloo, Ontario: Wilfrid Laurier University Press, 2006).

Wankmüller, Armin, 'Ausländische Studierende der Pharmazie und Chemie bei Liebig in Giessen', Deutsche Apotheker-Zeitung, 107/14 (1967), pp. 463–7.

Whewell, William, 'Review of "Art, III – On the Connexion of the Physical Sciences, by Mary Somerville"', Quarterly Review, 51 (March 1834), pp. 59–60.

White, Alan, Worcestershire Salt: A History of the Stoke Prior Salt Works (Bromsgrove: Halfshire Books, 1996).

White, Alan, The Worcestershire and Birmingham Canal: Chronicles of the Cut (Studley, Warwickshire: Brewin Books, 2005).

Whyte, Rebecca, 'Changing Approaches to Disinfection in England, c.1848–1914', PhD diss., University of Cambridge, 2012.

Wilmot, Sarah, 'Pollution and Public Concern: The Response of the Chemical Industry in Britain to Emerging Environmental Issues, 1860–1901', in Ernst Homburg, Anthony S. Travis and Harm G. Schröter (eds), The Chemical Industry in Europe, 1850–1914: Industrial Growth, Pollution and Professionalization, (Dordrecht: Kluwer Academic Publishers, 1998), pp. 121–47.

Wohl, Anthony S., Endangered Lives: Public Health in Victorian Britain (London: Methuen, 1974).

Woodin, S. and U. Skiba, 'Liming fails the acid test', New Scientist, (10 March 1990), pp. 30–34.

Woulfe, Peter, 'Experiments on the Distillation of Acids, volatile Alkalies, etc. showing how they may be condensed without Loss and how thereby we may avoid disagreeable and noxious Fumes: in a Letter from Mr. Peter Woulfe, F.R.S. to John Ellis, Esq; F.R.S.', Philosophical Transactions, 57 (1767), pp. 517–36.

Wynne, Brian, 'Establishing the rules of laws: constructing expert authority', in Roger Smith and Brian Wynne (eds), Expert Evidence: Interpreting Science and the Law (London: Routledge, 1989), pp. 23–55.

Index

Since the entire book is about Robert Angus Smith, the use of this name as an entry point has been restricted. Information will be found under the corresponding detailed topics.

For Product Safety Concerns and Information please contact our
EU representative GPSR@taylorandfrancis.com Taylor & Francis
Verlag GmbH, Kaufingerstraße 24, 80331 München, Germany